Chitosan and Other Edible Coatings with Antimicrobial Activity: Synthesis, Properties and Horticultural Applications

Chitosan and Other Edible Coatings with Antimicrobial Activity: Synthesis, Properties and Horticultural Applications

Editors

María B. Pérez-Gago
Lluís Palou

MDPI

Basel • Beijing • Wuhan • Barcelona • Belgrade • Novi Sad • Cluj • Manchester

Editors

María B. Pérez-Gago
Valencian Institute of
Agrarian Research (IVIA)
Valencia
Spain

Lluís Palou
Valencian Institute of
Agrarian Research (IVIA)
Valencia
Spain

Editorial Office
MDPI
St. Alban-Anlage 66
4052 Basel, Switzerland

This is a reprint of articles from the Special Issue published online in the open access journal *Coatings* (ISSN 2079-6412) (available at: https://www.mdpi.com/journal/coatings/special_issues/edible_coat_antimicrob).

For citation purposes, cite each article independently as indicated on the article page online and as indicated below:

Lastname, A.A.; Lastname, B.B. Article Title. *Journal Name* **Year**, *Volume Number*, Page Range.

ISBN 978-3-7258-0661-4 (Hbk)
ISBN 978-3-7258-0662-1 (PDF)
doi.org/10.3390/books978-3-7258-0662-1

Cover image courtesy of María B. Pérez Gago

Contents

coatings

MDPI

Article

Antifungal Starch–Gellan Edible Coatings with Thyme Essential Oil for the Postharvest Preservation of Apple and Persimmon

Mayra Sapper [1,*], Lluís Palou [2], María B. Pérez-Gago [2] and Amparo Chiralt [1]

1 Instituto Universitario de Ingeniería de Alimentos para el Desarrollo, Universitat Politècnica de València, Camino de Vera s/n, 46022 Valencia, Spain; dchiralt@tal.upv.es

2 Centro de Tecnología Poscosecha (CTP), Instituto Valenciano de Investigaciones Agrarias (IVIA), 46113 Moncada, Valencia, Spain; palou_llu@gva.es (L.P.); perez_mbe@gva.es (M.B.P.-G.)

* Correspondence: maysap@etsiamn.upv.es; Tel.: +34-644-323-055

Received: 9 May 2019; Accepted: 21 May 2019; Published: 24 May 2019

Abstract: Starch–gellan (80:20) coating formulations were applied to apples and persimmons to analyse their effectiveness controlling the weight loss, respiration rate, fruit firmness, and fungal decay during postharvest. Thyme essential oil (EO) was incorporated (0.25 and 0.5 g per g of polymer) directly or encapsulated in lecithin to enhance antifungal action. Coatings did not reduce weight loss or firmness changes in apples, but they prevented water loss in persimmons. In contrast, no significant effect of the coatings was observed on the respiration rates and the respiration quotient of persimmons, whereas the respiration rates and quotient was increased in apples. On the other hand, the coatings without lecithin reduced the incidence and severity of black spot caused by *Alternaria alternata* in persimmons, regardless of the content of essential oil. Likewise, these reduced the severity of gray mold caused by *Botrytis cinerea* in apples. No positive effect of lecithin in coatings was observed on the postharvest quality and decay in either apples or persimmons, nor did EO exert antifungal action despite its proven effectiveness in in vitro tests.

Keywords: cassava starch; gellan; thyme essential oil; antifungal edible coatings; postharvest; fruit

1. Introduction

Postharvest diseases are one of the major factors that affect the quality of horticultural fresh products during storage. Since fruit and vegetables are living organisms, their shelf life is greatly affected by temperature, relative humidity (RH), composition of the atmosphere during and after harvest, and the type and degree of infection by microorganisms or attack by insects [1]. Fruit contains high levels of sugars and nutrient elements, and the low pH values make them particularly susceptible to fungal decay. Fruit fungal infection may occur during flowering, fruit growth, harvesting, transport, packing operations, postharvest storage, or after purchase by the consumer [2]. Moreover, the natural resistance of fruit and vegetables to disease declines with storage duration and ripeness [3].

Gray mold caused by *Botrytis cinerea* Pers. is considered one of the most serious and common postharvest diseases of various fruit, including apples and persimmons [4,5]. The infection may occur in the field, from bloom to harvest, or after harvest, typically causing nests of decay. In fruit that are often stored for extended periods such as apples, field infections that remained latent can resume growth during storage, when the pathogen takes advantage of fruit maturity and environmental conditions and the disease develops (low temperatures and high humidity). In this sense, *B. cinerea* is very well adapted to low temperatures, and it is even able to grow at 0 °C [6]. Infection starts with a darker circular area where the fruit tissues are softer than the other fruit parts, and subsequent abundant sporification, whose colour ranges from white to gray, can develop from the site of infection in conditions of ambient temperature and high humidity [7]. *Alternaria alternata* (Fr.) Keissl. is the causal agent of postharvest black spot in persimmons (*Diospyros kaki* Thunb.) [8], and is generally considered a weak and opportunistic pathogen that gains entry into the fruit via wounds or natural openings, and remains quiescent until the fruit ripens [9]. *A. alternata* and *Penicillium* spp. were found to be the main causal agents of latent and wound infections in persimmons in Spain [5].

After harvest, fresh produce also suffer physiological and biochemical changes that cause detrimental changes in quality and shelf life. Respiration, transpiration, and ethylene production are the main factors contributing to the deterioration of fruits and vegetables [10]. Ethylene is a hormone produced by climacteric fruits, or when fruit undergoes stress, and is partially responsible for changes in the flavour, colour, and texture of fruits and vegetables. In addition, fresh fruits and vegetables lose water during storage due to respiratory and transpiration processes [11]. Water stress also causes metabolic alterations and changes in enzyme activation, causing accelerated senescence, a decline in nutritional value, and increased susceptibility to chilling injury and pathogen invasion. Respiration consists of the oxidative breakdown of organic reserves to simpler molecules, including carbon dioxide (CO_2) and water, with the release of energy [12]. All these biological factors, such as respiration, ethylene production, and resistance to water diffusion depend on the fruit commodity and cultivar, physiological stage at harvest, and storage conditions, which are also related to the composition of the surface waxes. Thus, for example, Morice and Shorland [13] reported that hydrocarbons, alcohols, fatty acids, ursolic acid, and α-farnesene are the main components in natural apple surface waxes, and the amount and composition of these components changed during storage depending on the apple cultivar.

In the last decade, considerable research has been carried out into the development of edible coatings aiming to control the physiological activity of fruit. These coatings can modify the internal gas composition and reduce the water loss through the regulation of oxygen (O_2), CO_2, and water vapour exchange between the fruit and the surrounding atmosphere. However, a certain degree of O_2 and CO_2 permeability is necessary to avoid anaerobic respiration, which induces ethanol production, off-flavour formation, and the loss of produce quality [10]. An additional advantage of edible coatings is the possibility of incorporating food-grade ingredients, such as antimicrobial agents, antioxidants, flavours, colour pigments, and vitamins into the basic formulation with the aim of improving their functional properties.

Traditionally, the postharvest disease control of fresh fruits and vegetables involves the use of synthetic chemical fungicides in those products for which their use is legislated. However, new restrictive regulations regarding fungicide residues, the reduction of the legal acceptability limits of specific fungicides, the emergence of fungicide-resistant strains of the pathogens, and an increasing public concern towards these compounds have led to a global increase in the need to seek safer postharvest alternatives to control the decay of fruits and vegetables [14]. Some of these include antimicrobial antagonists (bacteria, yeast, and fungi) that perform as biocontrol agents, synthetic and natural antimicrobials classified as food-grade additives, or generally recognized as safe (GRAS) compounds, such as organic and inorganic acids and their salts, chitosan, essential oils (EOs), or other plant extracts and different physical methods. Among the natural compounds, EOs and their components have been reported to suppress fungal growth, both in in vitro and in vivo studies. Thus, for example, tea tree, palmarosa and star anise EO vapours completely inhibited the in vitro germination of the apple pathogen *Penicillium expansum* L. [15]; *Melissa officinalis* EO was effective against *B. cinerea*, *P. expansum*, and *Rhizopus stolonifer* (Ehrenb.) Vuill. in in vitro studies [16]. *Pulicaria mauritanica* EO was effective against *Alternaria* sp., *P. expansum*, and *R. stolonifer* [17]. In in vivo studies, the addition of lemon EO enhanced the antifungal activity of chitosan against *B. cinerea* in strawberries [18]; garlic extracts and clove EO treatments reduced the postharvest decay caused by *B. cinerea* and *P. expansum* when applied directly to apples [19]; and a chitosan–oregano EO emulsion exhibited an inhibitory effect on pomegranate fruit inoculated with *Botrytis* sp., but caused some phytotoxicity [20]. Usually, the antibacterial effect of EOs relies on their high content of terpenes and terpenoids and also on the content of other aromatic and aliphatic constituents, all of which are characterized by low molecular weight [21].

However, in spite of the great potential of EOs, the main limitation to their application for decay control is the possible induction of a strong odour or flavour in fruit, phytotoxicity risks, and technological issues associated with commercial-scale fumigations or liquid applications [22]. The addition of EOs to edible coatings based on polymeric matrices could render them more effective at prolonging the postharvest life of horticultural produce, slowing down the diffusion rate of the antimicrobial agent and maintaining a higher concentration of the active compound on the fruit surface for a longer period, while preventing phytotoxicity by avoiding the direct contact of the EO with the fruit skin through the encapsulating action of the polymer. Additionally, fruit coatings can delay or retard the ripening process in climacteric fruit by modifying their internal gas composition and changing their permeability to O_2, CO_2, and ethylene production [23]. Among the different EOs, thyme EO exhibited antifungal action against *B. cinerea* and *A. alternata* when included in starch–gellan films in in vitro studies [24], as well as a complete growth inhibition of *B. cinerea* as vapour in in vitro tests [25]. Gellan, a microbial gum consisting of repeating tetrasaccharide units of glucose, glucuronic acid, and rhamnose residues joined in a linear chain, forms starch–gellan composite films with improved mechanical and barrier properties [24], which could be effective at preserving fruit quality during postharvest storage when applied as coatings. Furthermore, given the antifungal effect of these films when these contained thyme essential oil [24], their application as fruit coatings could represent a good strategy to extend the fruit postharvest life in terms of both quality maintenance and fungal growth inhibition. Nevertheless, in vivo assays in different fruits must be carried out in order to validate the beneficial effect of these coatings.

In this study, starch–gellan coatings incorporating thyme (*Thymus zygis* Loefl. ex L.) EO were applied to apples and persimmons to evaluate: (1) the postharvest behaviour of coated fruit in terms of weight loss, respiration rates, and mechanical properties, and (2) the antifungal efficacy of these coatings applied as a curative treatment against *B. cinerea* in apple and *A. alternata* in persimmon.

2. Materials and Methods

2.1. Reagents

To prepare the coating-forming systems (CFS), cassava starch (S) (with 10% amylose content) (Quimidroga S.A., Barcelona, Spain), low acyl gellan gum (G) (KELCOGEL F, Premium Ingredients, Murcia, Spain), non-GMO soy lecithin with 45% phosphatidylcholine (L) (Lipoid P45, Lipoid GmbH, Ludwigshafen, Germany) and thyme (*T. zygis*) essential oil (Plantis, Artesanía Agrícola SA, Barcelona, Spain) (EO) were used. The glycerol used as plasticizer was supplied by Panreac Química S.A. (Castellar de Vallès, Barcelona, Spain) and the polyoxyethylenesorbitan trioleate (Tween 85®) (T) was purchased from Sigma-Aldrich (Madrid, Spain).

2.2. Preparation of CFS

The formulations were prepared using S and G in a ratio of 8:2, with glycerol as the plasticiser (0.25 g per g of polymer), on the basis of previous studies [26,27]. Firstly, S was dispersed in distilled water and kept at 95 °C for 30 min to induce complete starch gelatinization. Meanwhile, G solution was obtained under stirring at 90 °C for 60 min. Both solutions were cooled down and afterwards, glycerol was added. The S and G systems were mixed to obtain the solutions without EO. The thyme EO (0.25 g per g and 0.5 g per g of polymer), which was used as an antifungal agent, was incorporated, either by direct emulsification or encapsulated in lecithin liposomes (polymer: lecithin ratio of 1:0.5). In the first case, the EO was added directly and the dispersions were homogenized for 3 min at 13.500 rpm using a rotor-stator homogenizer (Ultraturrax Yellow Line DL 25 Basic, IKA, Staufen, Germany). In the second case, the liposome dispersions were previously prepared and added directly to the initial polymer blend solution and kept under soft magnetic stirring for 2 h. A formulation was also obtained with lecithin liposomes without EO, as a control. To obtain the lecithin dispersions, lecithin (5%, $w \cdot w^{-1}$) was dispersed in distilled water and stirred for at least 4 h at 700 rpm. The EO (2.5% and 5% $w \cdot w^{-1}$) was added to the lecithin dispersion by using a sonicator (Vibra Cell, Sonics and Materials, Inc., Newtown, CT, USA) at 20 kHz for 10 min with pulses of 1 s, as described by Valencia-Sullca et al. [28]. Tween 85 was also added to S:G CFS (10^5 $mg \cdot L^{-1}$) in order to ensure the complete wettability of the fruit surface, according to a previous study [29], and tested on apples in a preliminary test. All the solutions were degassed using a vacuum pump (MZ 2C NT, Vacuubrand GmbH + CO KG, Germany). A total of six formulations were obtained: starch:gellan (S:G), control with lecithin (S:G-L), formulations with EO, non-encapsulated (S:G-0.25 and S:G-0.5), and lecithin-encapsulated (S:G-0.25-L and S:G-0.5-L).

2.3. *Rheological Behaviour and Contact Angle of the CFS*

The rheological behaviour was analysed in triplicate at 25 °C by means of a rotational rheometer (HAAKE Rheostress 1, Thermo Electric Corporation, Karlsruhe, Germany) by using a sensor system of coaxial cylinders, type Z34DIN Ti. Measurements were taken between 0–100 s^{-1}. The obtained data was fitted to the Ostwald de Waale power law model (Equation (1)) in order to determine the consistency (K) and the flow behaviour indices (n):

$$\sigma = K \cdot \left(\frac{\partial u}{\partial y} \right)^n \tag{1}$$

where σ = shear stress (Pa), K = flow consistency index ($Pa \cdot s^n$), $\frac{\partial_u}{\partial_y}$ = shear rate (s^{-1}), and n = the flow behaviour index.

The contact angle (θ) was determined by means of a Dynamic Contact Angle measuring device and Tensiometer (OCA 20, DataPhysics Instruments GmbH, Filderstadt, Germany). For this purpose, thin sections of the skin of the fruit were cut and placed on a glass plate to proceed with the measurements. Then, a droplet of each formulation was placed on the horizontal surface with a needle of 1.19 mm in internal diameter, and the contact angle at the fruit surfaces was measured by the sessile drop method [30]. Measurements were taken in less than 10 s. Image analyses were carried out using SCA20 software. At least 12 replicates were obtained.

2.4. *Quality of Coated Fruit*

Apples (*Malus domestica* Borkh cv. *Golden Delicious*) and persimmons (*Diospyros kaki* Thunb. cv. Rojo Brillante) were purchased from local packinghouses (Valencia, Spain) before any postharvest treatments were applied. Fruit were chosen according to their uniform shape, size, colour, and the absence of surface defects; then, they were subsequently cleaned and disinfected by a 4-min immersion in a 1% sodium hypochlorite solution, thoroughly rinsed with tap water, and air-dried at room temperature before coating application.

CFS were applied manually, using approximately 1.5 mL/fruit, and spread evenly over the fruit surface by using latex glove hands, following the method described by Bai et al. [31]. Water was applied to control fruit to simulate the coating application and its possible effect on the inoculum. Then, each fruit was inspected to assure complete coverage, and all fruit were stored at 25 °C and 65% RH, for 14 days. Ten fruit were considered in each series (coated and non-coated fruits).

2.4.1. Surface Density of Solids (SDS)

The SDS was determined by weighing the samples with a precision balance (Kern PFB 120-3, Germany) before and after coating application to obtain the CFS adhered mass. To calculate the total adhered solids, the mass fraction of each CFS was considered and the SDS ($g \cdot m^{-2}$) was estimated applying Equation (2), according to Marín et al. [32]:

$$SDS = \frac{(m_C - m_0) \cdot X_{sCFS}}{m_0} \cdot \rho \cdot \frac{1}{S_e} \quad (2)$$

where m_C = mass of the coated apple, m_0 = mass of the uncoated apple, X_{sCFS} = mass fraction of the solids of the CFS (g solids per g of solution), ρ = apple density ($g \cdot cm^{-3}$). To obtain the specific surface (S_e = 6/d, m^2 particles per m^3 fruit), the average diameter (d) was calculated considering a spherical geometry for the fruit.

2.4.2. Weight Loss

The weight loss of the fruit during storage was measured using an analytical balance (ME235P, Sartorius, Germany) before and after three, seven, and 14 days of storage. The mass loss was referred to the initial mass of the fruit, and the results were expressed as a relative mass loss rate (day^{-1}), which was obtained from the slope of the fitted straight line to the relative weight loss versus time data. Ten fruits were considered for each formulation and for control fruit.

2.4.3. Respiration Rates

Measurements were taken using a closed system, following the method proposed by Castelló et al. [33], with some modifications. Two apples were placed in hermetic glass jars with a septum in the lid for sampling headspace gas at different times. Gas sampling was carried out every 30 min for 4 h by means of a needle connected to a gas analyser (CheckMate 9900 PBI Dansensor, Ringsted, Denmark). Three replicates per treatment were performed after 7 and 14 days of storage. The respiration rate (R_i) of the samples in terms of CO_2 generation and O_2 consumption was determined from the slope of the fitted linear equation, according to Equation (3). The respiration quotient (RQ) has been determined as the ratio between CO_2 production and the O_2 consumption.

$$y_{it} = y_{i0} \pm 100 \cdot R_i \cdot \frac{M}{V} \cdot t \quad (3)$$

where y_{it} = gas concentration (%O_2, %CO_2) at time t, y_{i0} = initial gas concentration, R_i = respiration rate ($mL \cdot kg^{-1} \cdot h^{-1}$), M = mass of the samples, V = volume (mL) of headspace, and t = time.

2.4.4. Fruit Firmness

The firmness was measured using a Texture Analyser (Stable Micro Systems, TA.XT plus, Haslemere, England) fitted with an 11-mm diameter probe, applying a modification of the method described by Saei et al. [34]. A small skin area was removed from four opposite sides of each fruit around the equator. The probe penetrated the flesh at 10 mm min^{-1} and the maximum force (F_{max}, N) required to break the flesh was used as fruit firmness. The distance at maximum force (d_{max}, mm) was also taken as another representative parameter of the puncture curve. Ten replicates were used for each formulation after 14 days of storage. The same procedure was applied to uncoated fruit (control), both at the beginning and after 14 days of storage.

2.5. *In Vivo Antifungal Assays*

For the in vivo assays, *B. cinerea* strain BC03 from the IRTA Culture Collection (Lleida, Catalonia, Spain) was originally isolated from infected grapes from a vineyard located in Lleida and it was deposited at the Spanish Type Culture Collection (CECT-20973) at the University of Valencia (Burjassot, Valencia, Spain). *A. alternata* strain QAV-6 had been isolated from decayed persimmon fruit and maintained in the IVIA CTP Culture Collection of postharvest pathogens (Moncada, Valencia, Spain). These fungal strains were cultured on potato dextrose agar (PDA; Scharlab, Barcelona, Catalonia,

Spain) petri dishes at 25 °C in the dark and used after 7 to 14 days of active growth. Conidia were scraped from the cultures using a sterile loop and subsequently filtered and transferred to test tubes with sterile distilled water and 0.01% Tween 85. The suspensions were adjusted at 1×10^4 conidia mL^{-1} for *B. cinerea* and 5×10^5 conidia mL^{-1} for *A. alternata*, which were selected according to previous experience with these postharvest pathosystems [5,19]. The concentration of conidial suspensions was determined using a haemocytometer.

Fruit were wounded (approximately 1.6 mm in diameter and 2 mm deep) using the tip of a stainless-steel rod once in the fruit equator in the case of apples, and twice in the equator on the same side of the fruit in the case of persimmons (wounds located midway between the calyx and the stem end and 5–6 cm apart). Each wound was inoculated using a micropipette with 20 μL of the correspondent spore suspension 24 h before the application of the coatings (assessment of the coatings' curative activity). As previously described, coatings were applied manually at approximately 1.5 mL per fruit. Air surface drying was allowed at room temperature, and fruit were subsequently placed in perforated plastic trays avoiding direct contact between fruit and incubated at 20 °C and 85 ± 5% RH. Twenty fruit—four replications of five fruit each—were used per treatment. Control fruit were inoculated and treated with water using the same procedure as that for coating application. Lesion diameters (disease severity, mm) were measured after 7 and 12 days of incubation. Disease incidence (%) was expressed as the percentage of infected wounds out of the total number of inoculated wounds per replicate and treatment [6].

2.6. Statistical Analysis

The statistical analyses of the results were performed through an analysis of variance (ANOVA) using Statgraphics Centurion XVI.II (StatPoint Technologies Inc., Warrenton, VA, USA). Fisher's least significant difference (LSD) test was used at the 95% confidence level to determine specific differences between means. Multifactor ANOVA was also used to analyse the effect of the different factors (storage time and type of coating).

3. Results and Discussion

3.1. CFS Properties

The viscosity and contact angle of the different CFS on apple and persimmon skin were analysed since these parameters can affect the coating retention/adhesion on the fruit surface after the coating treatment through their influence on the CFS gravitational drainage before drying and liquid spreadability, all of which affect the coating thickness and homogeneity. The flow curves of the CFS were fitted to the power law model and the rheological parameters (consistency index: K and flow index: n), including the apparent viscosity (η) at 100 s^{-1}, are shown in Table 1. Pseudoplastic behaviour, with similar values of n—lower than 1—was observed in all the cases. Apparent viscosities ranged between 25–42 mPa·s, depending on the CFS composition. Directly-emulsified EO caused an increase in the apparent viscosity of the CFS according to the EO ratio, while lecithin-encapsulated EO reduced the viscosity of the formulations, which is probably due to the smaller droplet size in the encapsulated system [28]. Thus, the S:G-0.5 sample was the most viscous formulation and showed the highest consistency index.

The contact angles of the different CFS on apple and persimmon skin are also shown in Table 1. Values lower than 90° indicate surface wettability, and therefore greater extensibility of the coating on the fruit surface. For a given CFS, the contact angles on the persimmon skin were lower than on the apple skin, which indicates a better wettability of persimmon with these types of formulations. The values depended on the coating composition, with the highest contact angle corresponding to the S:G formulation in apples. This could imply problems for the extension of this coating on the apple surface. A previous study [29] reported that 10^5 mg·L^{-1} of Tween 85 must be added to ensure the S:G coating spreadability on the apple surface, whereas no surfactant was necessary to enhance the CFS spreadability when these contained emulsified or lecithin-encapsulated EO. Therefore, Tween 85 was added to the S:G formulation and tested in a preliminary trial with apples, in comparison with the CFS without Tween 85, in order to analyse the effect of the surfactant on the fruit quality during storage, as discussed in the next section.

Table 1. Rheological parameters (flow behavior index, n; consistency index, K; apparent viscosity at 100 s^{-1}, η) and contact angle (θ) of the coating forming solution (CFS) on the skin of 'Golden Delicious' apple and 'Rojo Brillante' persimmon. Mean values and standard deviations.

CFS	Rheological Behavior			Contact Angle (θ)	
	n	K (mPa·s) ^{n}	η at 100 s^{-1} (mPa·s)	Apple	Persimmon
S:G	0.854 ± 0.001 ^{d}	65.0 ± 0.2 ^{a}	33.1 ± 0.1 ^{c}	96 ± 2 ^{e}	67 ± 3 ^{cd}
S:G-L	0.74 ± 0.01 ^{a}	114 ± 3 ^{b}	35.0 ± 0.1 ^{d}	85 ± 3 ^{d}	72 ± 2 ^{e}
S:G-0.25	0.86 ± 0.01 ^{d}	59 ± 9 ^{a}	31 ± 3 ^{b}	69 ± 3 ^{a}	65 ± 3 ^{c}
S:G-0.25-L	0.815 ± 0.001 ^{c}	59.7 ± 0.5 ^{a}	25.5 ± 0.3 ^{a}	73 ± 2 ^{b}	50 ± 6 ^{a}
S:G-0.5	0.766 ± 0.004 ^{b}	124 ± 3 ^{c}	42.2 ± 0.2 ^{e}	77 ± 2 ^{c}	68 ± 2 ^{d}
S:G-0.5-L	0.809 ± 0.002 ^{c}	60 ± 1 ^{a}	25.05 ± 0.03 ^{a}	74 ± 2 ^{b}	55 ± 4 ^{b}

Different superscript letters within the same column indicate significant differences among CFS according to Fisher's least significant difference (LSD) test ($p < 0.05$).

3.2. *Effect of the Incorporation of Tween 85 into CFS on Apple Quality*

The incorporation of Tween 85 into the S:G CFS significantly decreased the contact angle on the apple surface (from 96 ± 2 to 47 ± 3) and increased the apparent viscosity (from 37.9 ± 0.3 to 187.8 ± 0.1 mPa·s). As expected, both changes affected the retention/adhesion of the CFS on the apple surface, as shown in the values of SDS on the fruit, which ranged from 2.6 ± 0.8 to 3.4 ± 0.5 g·m^{-2}. Interactions of Tween 85 with the CFS components and with the fruit surface affected both the viscosity and the contact angle of the CFS. As described by Marín et al. [32], surfactant molecules form complexes, with the helical conformation of amylose favouring the chain aggregation and increasing the system viscosity. Likewise, this complex formation implies that a high amount of surfactant is required to enhance the spreading of the CFS on the fruit surface, as discussed by Sapper et al. [29]. The increase in the SDS values for the S:G formulation with Tween 85 can be attributed to the higher solid content of the formulation, the greater viscosity that limits liquid gravitational drainage, and the lower contact angle. However, given the amphiphilic nature of the surfactant, its interactions with the natural wax of the fruit cuticle could also modify the overall barrier properties of the wax-coating assembly on the fruit surface. As is known, cuticular waxes are the primary components of the cuticle that are responsible for its permeability and wettability. These waxes are embedded in the cutin and form a continuous layer on the top of the cutin [35]. It has been reported that the cuticular wax content in apple fruit increases during fruit development and storage [36].

Table 2 shows the relative weight loss rate, respiration rate, and puncture parameters of apples after 7 days of storage at 25 °C for samples coated with the S:G formulation containing Tween 85 or not, in comparison with the uncoated control sample. Little differences in the relative weight loss rate were observed between the uncoated control sample and the one coated with the surfactant-free formulation. However, a significantly higher weight loss rate was observed for those coated with the formulation containing Tween 85. The coatings reduced the O_2 consumption rate of the fruit, which can be attributed to the low O_2 permeability of these films [24], but this reduction was particularly significant for the coating containing Tween 85. The CO_2 production rate was not significantly affected by the S:G coating compared to the control sample, but the coating containing Tween 85 significantly reduced this rate. As a consequence, the respiratory quotient was higher than 1 for both coated samples, indicating the creation of a modified atmosphere in the fruit and a shift towards anaerobic respiration pathways. The incorporation of Tween 85 resulted in a general decrease in the gas transfer rate and an increase in water transfer rate. As discussed above, the interactions of Tween 85 with the cuticular waxes, as well as its effect on the decrease in the cohesion forces of the S:G matrix (limiting of chain packing), could explain the changes observed in the gas and water vapour barrier properties of the coating and their effect on the fruit. These changes also had an effect on the fruit texture, as shown in Table 2. Although all the samples exhibited similar fruit firmness as deduced from the lack of significant differences among treatments regarding the maximum puncture force, there were significant differences in the maximum penetration distance (d_{max}) at the tissue rupture. Fruits coated with the S:G formulation containing Tween 85 had significantly higher d_{max} values, which reflect changes in the tissue texture. This fact can be related to the greater loss of water, and therefore, cellular turgidity, which is associated with a more marked superficial dehydration of the fruit with this coating. This factor is considered one of the main causes of texture changes in fruit [37]. After 7 days of storage, all the samples had higher d_{max} than the fruit at the initial time, with those coated with the CFS containing Tween 85 being significantly more deformable. Therefore, the use of Tween in the S:G formulation to improve its wettability on the apple surface was discarded on the basis of the negative effects on the fruit weight loss and texture.

Table 2. Effect of the incorporation of Tween 85 into the cassava starch:low acyl gellan gum (S:G) coating formulation on the postharvest behavior and quality of coated 'Golden Delicious' apples: relative weight loss rate (day^{-1}), respiration rates (consumption of O_2 and production of CO_2, $mL \cdot kg^{-1} \cdot h^{-1}$), respiration quotient (RQ), and values of the maximum puncture force (F_{max}, N) and penetration distance (d_{max}, mm) after 7 days of storage at 25 °C. Uncoated samples were used for values at harvest and the control after 7 days of storage.

	Control	Control	S:G	S:G-Tween 85
	Initial Time		**7 days**	
Weight loss rate	–	0.36 ± 0.02 [a]	0.36 ± 0.01 [a]	0.66 ± 0.06 [b]
F_{max}	43 ± 7	46 ± 6 [a]	49 ± 8 [a]	46 ± 4 [a]
d_{max}	3.0 ± 0.3	3.6 ± 0.4 [a]	3.9 ± 0.5 [a]	4.6 ± 0.7 [b]
R O_2	12.94 ± 0.05	12.9 ± 1.3 [b]	11.4 ± 0.8 [b]	7.77 ± 0.02 [a]
R CO_2	13.9 ± 0.6	15.2 ± 0.7 [b]	18 ± 1 [c]	11.0 ± 0.2 [a]
RQ	1.07 ± 0.05	1.18 ± 0.07 [a]	1.58 ± 0.03 [c]	1.41 ± 0.03 [b]

Different superscript letters within the same row indicate significant differences among CFS according to Fisher's LSD test ($p < 0.05$).

3.3. *Effect of CFS on Postharvest Behaviour and Quality of Apples and Persimmons*

Table 3 shows the initial values of respiration rates and puncture parameters of apples and persimmons, and Table 4 shows the same parameters, together with the values of SDS, for coated and uncoated 'Golden Delicious' apples and 'Rojo Brillante' persimmons after storage at 25 °C. SDS values are indicators of the coating thickness on the fruit; the higher the SDS, the thicker the coating. The SDS value depends on the amount of CFS that adhered to the surface of the fruit and the total solid content of the formulation. The former, in turn, is affected by the wetting/spreading capacity and the viscosity of the coating formulations. In general, the SDS values were higher in apples than in persimmons, which could be related to differences in both the surface tension of the skin [29] and in the skin morphology of the fruit. Thus, persimmons are characterized by a smooth skin, where the lack of small superficial pores could limit the capillary retention of the liquid fraction. Similarly, the CFS composition slightly affected the SDS differently depending on the fruit. In apples, the presence of lecithin in the formulation significantly reduced the SDS, whereas smaller differences associated with the CFS composition were observed in persimmons, and these were seemingly more closely related to the solid content of the CFS (incorporation of EO and/or lecithin to the formulations).

Table 3. Respiration rates (consumption of O_2 and production of CO_2, $mL \cdot kg^{-1} \cdot h^{-1}$), respiration quotient (RQ), and maximum puncture force (F_{max}, N) and penetration distance (d_{max}, mm) of uncoated 'Golden Delicious' apples and 'Rojo Brillante' persimmons at initial time. Mean values and standard deviations.

	R O_2	**R CO_2**	**RQ**	F_{max}	d_{max}
Apple	13.7 ± 1.5	12.2 ± 0.2	0.9 ± 0.1	29 ± 2	2.2 ± 0.1
Persimmon	5.7 ± 1.1	5.4 ± 0.9	0.94 ± 0.02	21.1 ± 0.5	2.5 ± 0.3

Table 4. Surface density of solids (SDS, $g \cdot m^{-2}$), weight loss rate (day^{-1}), respiration rates (consumption of O_2 and production of CO_2, $mL \cdot kg^{-1} \cdot h^{-1}$), respiration quotient (RQ), maximum puncture force (F_{max}, N), and penetration distance (d_{max}, mm) of coated and uncoated 'Golden Delicious' apples and 'Rojo Brillante' persimmons for 7 or 14 days of storage at 25 °C.

	Control	**S:G**	**S:G-L**	**S:G-0.25**	**S:G-0.25-L**	**S:G-0.5**	**S:G-0.5-L**
				Apple			
SSD	–	1.3 ± 0.3^{c}	0.8 ± 0.1^{a}	1.4 ± 0.3^{c}	1.1 ± 0.1^{b}	1.5 ± 0.2^{c}	1.0 ± 0.1^{b}
Weight loss rate (14 d)	0.23 ± 0.03^{a}	0.21 ± 0.03^{a}	0.20 ± 0.05^{a}	0.21 ± 0.03^{a}	0.20 ± 0.03^{a}	0.20 ± 0.03^{a}	0.22 ± 0.03^{a}
R O_2 (7 d)	6.0 ± 1.4^{ab}	6.0 ± 1.5^{ab}	6.5 ± 1.6^{ab}	6.6 ± 0.8^{ab}	6.5 ± 1.0^{ab}	7.6 ± 0.6^{b}	5.1 ± 0.9^{a}
R CO_2 (7 d)	6.5 ± 1.2^{a}	7.8 ± 0.8^{ab}	8.4 ± 2.0^{ab}	8.8 ± 0.1^{ab}	8.3 ± 0.7^{ab}	9.7 ± 1.0^{b}	6.8 ± 0.9^{a}
RQ (7 d)	1.1 ± 0.1^{a}	1.3 ± 0.1^{b}	1.29 ± 0.04^{b}	1.4 ± 0.1^{b}	1.3 ± 0.1^{b}	1.3 ± 0.1^{b}	1.3 ± 0.1^{b}
R O_2 (14 d)	5.3 ± 0.8^{a}	8.4 ± 1.1^{d}	7.3 ± 0.5^{cd}	7.0 ± 0.1^{bcd}	5.9 ± 0.8^{ab}	6.5 ± 0.9^{abc}	5.6 ± 0.9^{ab}
R CO_2 (14 d)	6.2 ± 1.3^{a}	10.4 ± 1.1^{d}	9.0 ± 0.9^{cd}	8.1 ± 0.2^{bc}	7.2 ± 0.6^{ab}	7.9 ± 0.6^{bc}	7.2 ± 0.6^{ab}
RQ (14 d)	1.2 ± 0.1^{a}	1.2 ± 0.1^{a}	1.23 ± 0.05^{a}	1.2 ± 0.1^{a}	1.2 ± 0.1^{a}	1.2 ± 0.1^{a}	1.3 ± 0.1^{a}
F_{max} (14 d)	27 ± 2^{ab}	31 ± 3^{c}	30 ± 4^{bc}	29 ± 3^{abc}	30.4 ± 1.5^{bc}	32 ± 2^{c}	26.1 ± 1.2^{a}
d_{max} (14 d)	3 ± 1^{a}	3 ± 1^{a}	3 ± 1^{a}	3 ± 0^{a}	3 ± 1^{a}	3 ± 1^{a}	3.4 ± 0.5^{a}
				Persimmon			
SSD	–	0.5 ± 0.1^{a}	0.7 ± 0.1^{ab}	0.7 ± 0.2^{b}	0.8 ± 0.1^{bc}	0.9 ± 0.1^{c}	0.7 ± 0.2^{b}
Weight loss rate (14 d)	0.7 ± 0.1^{b}	0.6 ± 0.1^{ab}	0.6 ± 0.1^{ab}	0.52 ± 0.03^{a}	0.56 ± 0.06^{ab}	0.6 ± 0.1^{ab}	0.7 ± 0.1^{b}
R O_2 (7 d)	5.2 ± 0.3^{a}	3.3 ± 1.6^{a}	3.8 ± 0.6^{a}	2.9 ± 0.2^{a}	3.9 ± 2.1^{a}	3.6 ± 0.3^{a}	4.1 ± 1.7^{a}
R CO_2 (7 d)	5.6 ± 0.5^{a}	3.8 ± 1.0^{a}	4.2 ± 0.9^{a}	3.4 ± 0.4^{a}	4.5 ± 2.0^{a}	4.6 ± 0.7^{a}	4.8 ± 2.1^{a}
RQ (7 d)	1.08 ± 0.04^{a}	1.2 ± 0.3^{a}	1.11 ± 0.05^{a}	1.19 ± 0.03^{a}	1.3 ± 0.3^{a}	1.3 ± 0.1^{a}	1.2 ± 0.1^{a}
R O_2 (14 d)	3.7 ± 0.6^{a}	4.2 ± 1.8^{a}	3.0 ± 1.1^{a}	3.3 ± 0.4^{a}	2.2 ± 0.2^{a}	2.6 ± 0.6^{a}	3.5 ± 1.9^{a}
R CO_2 (14 d)	3.9 ± 1.1^{a}	5.3 ± 2.6^{a}	4.1 ± 1.0^{a}	3.9 ± 1.8^{a}	2.2 ± 0.2^{a}	2.9 ± 0.4^{a}	4.2 ± 1.3^{a}
RQ (14 d)	1.0 ± 0.1^{a}	1.3 ± 0.1^{a}	1.4 ± 0.2^{a}	1.2 ± 0.4^{a}	1.0 ± 0.0^{a}	1.1 ± 0.1^{a}	1.3 ± 0.3^{a}
F_{max} (14 d)	21 ± 4^{a}	27 ± 3^{bc}	23 ± 4^{ab}	27 ± 5^{bc}	28 ± 6^{c}	29 ± 6^{c}	28 ± 6^{c}
d_{max} (14 d)	5 ± 1^{ab}	6 ± 1^{bc}	5 ± 1^{a}	5 ± 1^{a}	5.2 ± 0.4^{abc}	6 ± 1^{c}	6 ± 1^{bc}

Different superscript letters within the same row indicate significant differences among CFS according to Fisher's LSD test ($p < 0.05$).

The rate of relative weight loss after 14 days of storage was not significantly affected by coating application or composition, and ranged between 0.20–0.23 day^{-1}. In persimmons, water loss rates were higher than in apples and varied depending on the coating formulation. The highest values (0.7 day^{-1}) were obtained for uncoated samples, and those coated with CFS containing the highest content of lecithin-encapsulated EO (maximum lipid content in the film) and the lowest value (0.52 day^{-1}) was obtained for samples coated with the formulation with emulsified EO (without lecithin) at the lowest ratio (minimum lipid content in the film). No significant differences were found between the other coating formulations and the control samples. These results indicate that persimmon fruits were more sensitive to dehydration than apples under these storage conditions, and the coatings with the lowest ratio of emulsified EO exerted a protective effect. As mentioned above, apples and persimmons are naturally covered by a continuous wax layer that provides the resistance to water movement across the cuticle. The differences in the water resistance of the untreated fruits can be attributed to the particular fruit physiology, skin morphology, and the composition of the natural waxes. The application of coatings containing hydrophobic compounds should improve the moisture resistance of the fruit, as an additional layer is deposited over the natural waxes. In the present study, none of the coatings reduced weight loss in apples, and only the coating that had the lowest amount of EO and no lecithin prevented water loss in persimmons. This might indicate a partial removal and/or modification of the natural waxes that are present on the peel of the fruits, resulting in no reduction in weight loss; so, further studies should be conducted in order to understand the effect of the EO and lecithin on the water barrier properties of coated apples and persimmons. Some other studies also reflected no effect of coatings based on biopolymers and lipids on the weight loss reduction of different fruits, such as apples [31], plums [38], table grapes [39], or cherry tomatoes [40], compared to uncoated fruits.

Table 4 also shows respiration rates of both fruit at 7 and 14 days of storage. In apples, a multifactor ANOVA (results not shown) did not reveal a significant effect of the storage time on respiration rates, although the coatings had a significant influence. In general, coatings tend to increase the O_2 consumption and CO_2 production rates with respect to the control sample, and only those containing lecithin-encapsulated EO showed no significant differences with respect to the control sample. The alterations in the respiration pathway affected the RQ, which indicates the nature of the substrate used during the respiration process. Thus, a RQ equal to 1.0 indicates that the metabolic substrates are carbohydrates, whereas an RQ higher than 1 indicates that the substrates are organic acids [41]. The multifactor ANOVA in RQ reveals a significant effect of storage time and coating formulation. RQ slightly increased at 14 days, and was higher in all the coated samples. This indicates that the metabolic substrates are shifting from carbohydrates to organic acids [41] more quickly in coated samples.

In contrast, coatings were observed to have no significant effect on the respiration rates of persimmons, which exhibited lower respiration rates than apples, with a respiration quotient of nearly 1. Climacteric fruit, such as apples, exhibit a peak of respiration and ethylene (C_2H_4) production associated with senescence or ripening [12], which could explain the observed differences.

The texture changes in fruits depend on both cell wall degradation and the loss of tissue turgidity [42]. Table 4 shows the values of the F_{max} and d_{max} for the different coated and uncoated fruits after 14 days of storage. No significant changes in the maximum puncture force (failure point) were observed for either uncoated or coated apples after storage with respect to the initial values before storage (Tables 3 and 4). However, although no significant differences were observed in terms of the maximum penetration distance between coated and control apples at the end of the storage, the values were slightly higher than before storage, which can be associated with a loss of cellular turgidity due to the superficial dehydration of the apples. The limited water vapor barrier capacity of these films [24] and their relative lack of thickness on the fruit mean that these are scarcely effective at controlling moisture transfer in apples.

In the case of persimmons, the coatings had a significant effect ($p < 0.05$), maintaining the firmness of the fruit. On the other hand, although there were no notable differences in terms of the maximum penetration distance at the failure point (5–6 mm) between coated and uncoated samples at the end of the storage, the values were significantly higher than the initial value (2.5 mm) before storage. This indicates changes in the texture of the tissue over time, which can be related to the progress in maturity and water loss. The F_{max} values increased from 21.1 up to 31 N, which could be attributed to the greater deformability of the tissue allowing for deeper penetration without failure, thus accumulating more compressive and shear resistance [43]. The smallest changes occurred in the sample coated with CFS containing lecithin without EO.

The above results show that coatings have a different effect depending on whether the fruit is an apple or persimmon, which can be attributed to the different physiological patterns of the fruits and the specific interactions with the coatings. Although respiration patterns were slightly modified by coatings on apples with no effect on water loss, coatings exerted a better control of water loss in persimmon; however, these did not maintain the firmness mainly due to the progress of fruit ripening.

3.4. Fungal Decay

Table 5 shows the development of fungal decay on artificially inoculated 'Golden Delicious' apples and 'Rojo Brillante' persimmons. The applications of starch–gellan coatings did not significantly reduce the disease incidence on apples inoculated with *B. cinerea*, as compared to non-coated ones (control) after 7 or 12 days of storage at 20 °C. No effect of the addition of EO was observed, despite what had been observed in a prior in vitro study, where starch–gellan films with thyme EO exhibited a marked antifungal effect [24]. Nevertheless, all the coatings, regardless of their composition, significantly reduced the severity of gray mold with respect to the control samples (20–30% reduction), with no particular observed effect of the antifungal EO.

Table 5. Mean values and standard deviations of disease incidence and severity of gray mold on 'Golden Delicious' apples artificially inoculated with *Botrytis cinerea* and black spot on 'Rojo Brillante' persimmons artificially inoculated with *Alternaria alternata*. Fruit were coated 24 h after fungal inoculation and incubated at 20 °C and 85% RH for 7 and 12 days. Mean values of the reduction in disease incidence and severity are also shown.

	Disease Incidence (%)		**Reduction of Incidence (%)**		**Disease Severity (mm)**		**Reduction of Severity (%)**	
	7 Days	**12 Days**	**7 Days**	**12 Days**	**7 Days**	**12 Days**	**7 Days**	**12 Days**
				Apple gray mold				
Control	100 ± 0 ^{a}	100 ± 0 ^{a}	–	–	70 ± 5 ^{b}	100 ± 5 ^{b}	–	–
S:G	75 ± 25 ^{a}	83 ± 14 ^{a}	25	17	44 ± 5 ^{a}	74 ± 13 ^{a}	32	26
S:G-L	92 ± 14 ^{a}	92 ± 14 ^{a}	8	8	47 ± 11 ^{a}	73 ± 21 ^{a}	27	27
S:G-0.25	75 ± 25 ^{a}	83 ± 14 ^{a}	25	17	53 ± 8 ^{ab}	76 ± 15 ^{a}	19	24
S:G-0.25-L	75 ± 25 ^{a}	75 ± 25 ^{a}	25	25	44 ± 11 ^{a}	64 ± 17 ^{a}	33	36
S:G-0.5	92 ± 14 ^{a}	92 ± 14 ^{a}	8	8	45 ± 10 ^{a}	81 ± 7 ^{ab}	32	19
S:G-0.5-L	100 ± 0 ^{a}	100 ± 0 ^{a}	0	0	47 ± 1 ^{a}	69 ± 10 ^{a}	29	31
				Persimmon black spot				
Control	68 ± 3 ^{b}	73 ± 5 ^{bc}	–	–	10.6 ± 0.8 ^{a}	21.9 ± 1.9 ^{b}	–	–
S:G	38 ± 9 ^{a}	45 ± 12 ^{a}	44	39	9.3 ± 1.5 ^{a}	15.0 ± 3.0 ^{a}	12	32.9
S:G-L	70 ± 10 ^{b}	78 ± 7 ^{bc}	0	0	9.7 ± 1.2 ^{a}	17.0 ± 3.0 ^{a}	9	20.8
S:G-0.25	42 ± 7 ^{a}	57 ± 10 ^{ab}	39	23	10.8 ± 1.1 ^{a}	15.5 ± 1.4 ^{a}	0	29.4
S:G-0.25-L	58 ± 8 ^{ab}	72 ± 6 ^{bc}	14	2	12.7 ± 0.6 ^{a}	18.8 ± 1.9 ^{ab}	0	14.2
S:G-0.5	42 ± 7 ^{a}	53 ± 7 ^{ab}	39	27	11.4 ± 0.3 ^{a}	20.2 ± 0.5 ^{ab}	0	8
S:G-0.5-L	72 ± 7 ^{b}	82 ± 8 ^{c}	0	0	10.4 ± 0.8 ^{a}	17.6 ± 0.6 ^{a}	2	20

For each disease, different superscript letters within the same column indicate significant differences among CFS according to Fisher's LSD test ($p < 0.05$).

Starch–gellan coatings were more effective at reducing the incidence of black spot caused by *A. alternata* on persimmons (up to 40% reduction), although coatings containing lecithin were not effective, and the presence of EO was not observed to have any significant effect. Disease severity was not significantly reduced in coated persimmons. A multifactorial analysis (factors: the presence of lecithin and EO concentration) revealed two things: there was no significant influence of the EO, and the lecithin had a negative effect on the reduction of disease incidence and severity in infected fruit.

In general, applying a coating had a positive antifungal effect both on apples (a significant reduction in the severity of gray mold) and persimmons (a significant reduction in the incidence of black spot), but this antifungal effect was milder than that observed in in vitro work with EO incorporated into the same type of films. Similar behavior has recently been reported by da Rocha Neto et al. [15] for apples. They observed a complete inhibition of the in vitro germination of *P. expansum* by using melaleuca, palmarosa, and star anise EOs in vapor phase, but these treatments had only a minor effect on inoculated apples, regardless of the EO used. As previously reported [22,44], this indicates that the in vivo effectiveness of EOs cannot be anticipated by their antifungal activity in in vitro tests, and that interactions between EOs and fungal pathogens are modulated by the fruit host and the conditions in the infection court, often resulting in reduced disease control ability. An important difference in the potential effect of EO with respect to in vitro tests could be related with the degree of coating plasticization, which may affect the release of EO. In in vitro tests, films are directly applied on the wet culture medium, whereas coatings are applied on the dried fruit surface. This fact could limit the release of the active compounds from the polymer matrix, hindering their antifungal action.

Likewise, EO compounds may also affect some physiological changes in the fruit, which could decrease the fruit's natural defenses against the fungal attack. The generally negative effect of lecithin could also be attributed to the lipid interactions with the fruit's waxy coatings, which could also weaken the natural resistance to disease, counteracting the induced coating protection. Other surfactant lipids, such as Tween 85, also seemed to exert a negative effect on the barrier capacity of the natural wax-coating assembly, as observed in apples. The gas exchange on the fruit surface could also play an important role in postharvest disease development, which could explain the generally positive effect of coatings at reducing fungal growth on infected fruit. Therefore, interactions of coatings and their components with the fruit surface always constitute a distinguishing factor to define the particular behavior of coated fruit [45], and in the case of coatings formulated with antifungal ingredients, these interactions can affect the in vivo disease control ability of the coating [46]. The results obtained in the present study with the addition of thyme EO to starch-based coatings were not anticipated. Numerous previous studies have shown that the formulation of antifungal films and coatings with EOs either provided disease control ability or increased that of the coating alone due to an important synergistic effect against various important postharvest pathogens, including *B. cinerea* and *A. alternata* [12,47,48]. However, this is not always the case, and other reports showed no significant benefit gained from the addition of EOs [49]. It seems that a wide variability in disease control efficacy can be observed, which is basically due to the numerous factors that can influence the antifungal properties of films and coatings. The following can be cited among the most important: nature of the composite matrix of the coating; type and concentration of the antifungal compound(s); species and strain of the target postharvest pathogen; species, cultivar, and physical and physiological condition of the fruit host; and postharvest environmental conditions [46]. Therefore, further studies would be required to analyze the influence of these factors.

4. Conclusions

Starch–gellan coatings containing or not emulsified or lecithin-encapsulated EO had a different effect on the postharvest parameters (weight loss, respiration rates and firmness changes) when applied on apples and persimmons, depending on the coating composition and type of fruit. None of the coating formulations reduced the weight loss in apples, although these prevented water loss in persimmons. In contrast, although the coating was not observed to have any significant effect on the respiration rates and respiration quotient of persimmons, the respiration rates and quotient in apples were promoted. Coatings did not affect the changes in fruit firmness in apples or persimmons; nevertheless, in the latter, these may be mainly associated with the ripening progress. Regarding fungal decay, coatings without lecithin reduced the incidence of black spot caused by *A. alternata* in persimmons, regardless of the thyme EO content. Likewise, these reduced the severity of gray mold caused by *B. cinerea* infection in apple. The addition of EO did not exert an antifungal effect in the fruit despite its proven antifungal action in previous in vitro tests. Therefore, the particular characteristics of the fruit and the interactions in the infection site (peel wounds) seriously affected the in vivo effectiveness of coatings of certain composition. No positive effect of lecithin was observed on the controlled postharvest parameters affecting fruit quality and physiological behavior in either apples or persimmons; EO did not exert additional antifungal action and seemed to exert a negative effect on some other fruit quality attributes. Then, starch–gellan coatings without lecithin or thyme EO demonstrated the potential to be used in persimmons in order to control weight loss and reduce the incidence of infections caused by *A. alternata*.

Author Contributions: Conceptualization, M.S., L.P., M.B.P.-G., and A.C.; methodology, L.P., M.B.P.-G., and A.C.; investigation, M.S.; writing—original draft preparation, M.S. and A.C.; writing—review and editing, M.S., L.P., M.B.P.-G. and A.C.

Funding: This research was funded by the "Ministerio de Economía y Competitividad" (MINECO) of Spain through the projects AGL2016-76699-R and RTA2015-00037-C02. Mayra Sapper thanks the "Conselleria de Educación, Investigación, Cultura y Deporte de la Generalitat Valenciana" for the Santiago Grisolía grant GRISOLIA/2015/001.

Conflicts of Interest: The authors declare no conflict of interest.

References

1. Singh, D.; Sharma, R.R. Postharvest diseases of fruits and vegetables and their management. In *Postharvest Disinfection of Fruits and Vegetables*; Siddiqui, M.W., Ed.; Academic Press: London, UK, 2018; pp. 1–52. [CrossRef]
2. Sivakumar, D.; Bautista-Baños, S. A review on the use of essential oils for postharvest decay control and maintenance of fruit quality during storage. *Crop Prot.* **2014**, *64*, 27–37. [CrossRef]
3. Troncoso-Rojas, R.; Tiznado-Hernández, M.E. Alternaria alternata (black rot, black spot). In *Postharvest Decay: Control Strategies*; Bautista-Baños, S., Ed.; Academic Press: London, UK, 2014; pp. 147–187. [CrossRef]
4. Batta, Y.A. Postharvest biological control of apple gray mold by *Trichoderma harzianum* Rifai formulated in an invert emulsion. *Crop Prot.* **2004**, *23*, 19–26. [CrossRef]
5. Palou, L.; Montesinos-Herrero, C.; Tarazona, I.; Besada, C.; Taberner, V. Incidence and etiology of postharvest fungal diseases of persimmon (*Diospyros Kaki* Thunb. Cv. Rojo Brillante) in Spain. *Plant Dis.* **2015**, *99*, 1416–1425. [CrossRef] [PubMed]
6. Ma, L.; He, J.; Liu, H.; Zhou, H. The phenylpropanoid pathway affects apple fruit resistance to *Botrytis cinerea*. *J. Phytopathol.* **2018**, *166*, 206–215. [CrossRef]
7. Romanazzi, G.; Feliziani, E. Botrytis cinerea (gray mold). In *Postharvest Decay: Control Strategies*; Bautista-Baños, S., Ed.; Academic Press: London, UK, 2014; pp. 131–146. [CrossRef]
8. Prusky, D.; Eshel, D.; Kobiler, I.; Yakoby, N.; Beno-Moualem, D.; Ackerman, M.; Zuthji, Y.; Ben Arie, R. Postharvest chlorine treatments for the control of the persimmon black spot disease caused by *Alternaria alternata*. *Postharvest Biol. Technol.* **2001**, *22*, 271–277. [CrossRef]
9. Biton, E.; Kobiler, I.; Feygenberg, O.; Yaari, M.; Kaplunov, T.; Ackerman, M.; Friedman, H.; Prusky, D. The mechanism of differential susceptibility to alternaria black spot, caused by *Alternaria alternata*, of stem-and bottom-end tissues of persimmon fruit. *Postharvest Biol. Technol.* **2014**, *94*, 74–81. [CrossRef]
10. Olivas, G.; Barbosa-Cánovas, G. Edible films and coatings for fruits and vegetables. In *Edible Films and Coatings for Food Applications*; Embuscado, M.E., Huber, K.C., Eds.; Springer: New York, NY, USA, 2009; pp. 211–244. [CrossRef]
11. Maftoonazad, N.; Ramaswamy, H.S.; Marcotte, M. Shelf-life extension of peaches through sodium alginate and methyl cellulose edible coatings. *Int. J. Food Sci. Technol.* **2008**, *43*, 951–957. [CrossRef]
12. Fonseca, S.C.; Oliveira, F.A.R.; Brecht, J.K. Modelling respiration rate of fresh fruits and vegetables for modified atmosphere packages: A review. *J. Food Eng.* **2002**, *52*, 99–119. [CrossRef]
13. Morice, I.M.; Shorland, F.B. Composition of the surface waxes of apple fruits and changes during storage. *J. Sci. Food Agric.* **1973**, *24*, 1331–1339. [CrossRef]
14. Prusky, D.; Kobiler, I.; Akerman, M.; Miyara, I. Effect of acidic solutions and acidic prochloraz on the control of postharvest decay caused by *Alternaria alternata* in mango and persimmon fruit. *Postharvest Biol. Technol.* **2006**, *42*, 134–141. [CrossRef]
15. Da Rocha Neto, A.C.; Navarro, B.B.; Canton, L.; Maraschin, M.; Di Piero, R.M. Antifungal activity of palmarosa (*Cymbopogon martinii*), tea tree (*Melaleuca alternifolia*) and star anise (*Illicium verum*) essential oils against *Penicillium expansum* and their mechanisms of action. *LWT-Food Sci. Technol.* **2019**, *105*, 385–392. [CrossRef]
16. El Ouadi, Y.; Manssouri, M.; Bouyanzer, A.; Majidi, L.; Bendaif, H.; Elmsellem, H.; Shariati, M.A.; Melhaoui, A.; Hammouti, B. Essential oil composition and antifungal activity of *Melissa officinalis* originating from North-Est Morocco, against postharvest phytopathogenic fungi in apples. *Microb. Pathog.* **2017**, *107*, 321–326. [CrossRef] [PubMed]
17. Desjobert, J.M.; Cristofari, G.; Paolini, J.; Costa, J.; Majidi, L.; Znini, M. Essential oil composition and antifungal activity of *Pulicaria mauritanica* coss., against postharvest phytopathogenic fungi in apples. *LWT-Food Sci. Technol.* **2013**, *54*, 564–569. [CrossRef]

18. Perdones, A.; Sánchez-González, L.; Chiralt, A.; Vargas, M. Effect of chitosan-lemon essential oil coatings on storage-keeping quality of strawberry. *Postharvest Biol. Technol.* **2012**, *70*, 32–41. [CrossRef]
19. Daniel, C.K.; Lennox, C.L.; Vries, F.A. In vivo application of garlic extracts in combination with clove oil to prevent postharvest decay caused by *Botrytis cinerea*, *Penicillium expansum* and *Neofabraea alba* on apples. *Postharvest Biol. Technol.* **2014**, *99*, 88–92. [CrossRef]
20. Munhuweyi, K.; Caleb, O.J.; Lennox, C.L.; van Reenen, A.J.; Opara, U.L. In vitro and in vivo antifungal activity of chitosan-essential oils against pomegranate fruit pathogens. *Postharvest Biol. Technol.* **2017**, *129*, 9–22. [CrossRef]
21. Bakkali, F.; Averbeck, S.; Averbeck, D.; Idaomar, M. Biological effects of essential oils-A review. *Food Chem. Toxicol.* **2008**, *46*, 446–475. [CrossRef]
22. Palou, L.; Ali, A.; Fallik, E.; Romanazzi, G. GRAS, plant-and animal-derived compounds as alternatives to conventional fungicides for the control of postharvest diseases of fresh horticultural produce. *Postharvest Biol. Technol.* **2016**, *122*, 41–52. [CrossRef]
23. Sánchez-González, L.; Vargas, M.; González-Martínez, C.; Chiralt, A.; Cháfer, M. Use of essential oils in bioactive edible coatings: A review. *Food Eng. Rev.* **2011**, *3*, 1–16. [CrossRef]
24. Sapper, M.; Wilcaso, P.; Santamarina, M.P.; Roselló, J.; Chiralt, A. Antifungal and functional properties of starch-gellan films containing thyme (*Thymus zygis*) essential oil. *Food Control.* **2018**, *92*, 505–515. [CrossRef]
25. Plotto, A.; Roberts, D.D.; Roberts, R.G. Evaluation of plant essential oils as natural postharvest disease control of tomato (*Lycopersicon esculentum*). *Acta Hortic.* **2003**, *628*, 737–745. [CrossRef]
26. Cano, A.; Fortunati, E.; Cháfer, M.; Kenny, J.M.; Chiralt, A.; González-Martínez, C. Properties and ageing behaviour of pea starch films as affected by blend with poly(vinyl alcohol). *Food Hydrocoll.* **2015**, *48*, 84–93. [CrossRef]
27. Jiménez, A.; Fabra, M.J.; Talens, P.; Chiralt, A. Edible and biodegradable starch films: A review. *Food Bioprocess Technol.* **2012**, *5*, 2058–2076. [CrossRef]
28. Valencia-Sullca, C.; Jiménez, M.; Jiménez, A.; Atarés, L.; Vargas, M.; Chiralt, A. Influence of liposome encapsulated essential oils on properties of chitosan films. *Polym. Int.* **2016**, *65*, 979–987. [CrossRef]
29. Sapper, M.; Bonet, M.; Chiralt, A. Wettability of starch-gellan coatings on fruits, as affected by the incorporation of essential oil and/or surfactants. *LWT-Food Sci. Technol.* under review.
30. Kwok, D.Y.; Neumann, A.W. Contact angle measurement and contact angle interpretation. *Adv. Colloid Interface Sci.* **1999**, *81*, 167–249. [CrossRef]
31. Bai, J.; Baldwin, E.A.; Hagenmaier, R.H. Alternatives to shellac coatings provide comparable gloss, internal gas modification, and quality for "delicious" apple fruit. *HortScience* **2002**, *37*, 559–563. [CrossRef]
32. Marín, A.; Atarés, L.; Cháfer, M.; Chiralt, A. Properties of biopolymer dispersions and films used as carriers of the biocontrol agent *Candida sake* CPA-1. *LWT-Food Sci. Technol.* **2017**, *79*, 60–69. [CrossRef]
33. Castelló, M.L.; Fito, P.J.; Chiralt, A. Changes in respiration rate and physical properties of strawberries due to osmotic dehydration and storage. *J. Food Eng.* **2010**, *97*, 64–71. [CrossRef]
34. Saei, A.; Tustin, D.S.; Zamani, Z.; Talaie, A.; Hall, A.J. Cropping effects on the loss of apple fruit firmness during storage: the relationship between texture retention and fruit dry matter concentration. *Sci. Hortic.* **2011**, *130*, 256–265. [CrossRef]
35. Belding, R.; Blankenship, S.; Young, E.; Leidy, R. Composition and variability of epicuticular waxes in apple cultivars. *J. Am. Soc. Hortic. Sci.* **1998**, *123*, 348–356. [CrossRef]
36. Ju, Z.; Bramlage, W.J. Developmental changes of cuticular constituents and their association with ethylene during fruit ripening in "Delicious" apples. *Postharvest Biol. Technol.* **2001**, *21*, 257–263. [CrossRef]
37. Saberi, B.; Golding, J.B.; Marques, J.R.; Pristijono, P.; Chockchaisawasdee, S.; Scarlett, C.J.; Stathopoulos, C.E. Application of biocomposite edible coatings based on pea starch and guar gum on quality, storability and shelf life of 'Valencia' oranges. *Postharvest Biol. Technol.* **2018**, *137*, 9–20. [CrossRef]
38. Navarro-Tarazaga, M.L.; Sothornvit, R.; Pérez-Gago, M.B. Effect of plasticizer type and amount on hydroxypropyl methylcellulose-beeswax edible film properties and postharvest quality of coated plums (cv. Angeleno). *J. Agric. Food Chem.* **2008**, *56*, 9502–9509. [CrossRef] [PubMed]
39. Pastor, C.; Sanchez-Gonzalez, L.; Marcilla, A.; Chiralt, A.; Cháfer, M.; Gonzalez-Martinez, C. Quality and safety of table grapes coated with hydroxypropylmethylcellulose edible coatings containing propolis extract. *Postharvest Biol. Technol.* **2011**, *60*, 64–70. [CrossRef]

40. Fagundes, C.; Palou, L.; Monteiro, A.R.; Pérez-Gago, M.B. Hydroxypropyl methylcellulose-beeswax edible coatings formulated with antifungal food additives to reduce Alternaria black spot and maintain postharvest quality of cold-stored cherry tomatoes. *Sci. Hortic. (Amsterdam)* **2015**, *193*, 249–257. [CrossRef]
41. Fagundes, C.; Carciofi, B.A.M.; Monteiro, A.R. Estimate of respiration rate and physicochemical changes of fresh-cut apples stored under different temperatures. *Food Sci. Technol.* **2013**, *33*, 60–67. [CrossRef]
42. Ribeiro, C.; Vicente, A.A.; Teixeira, J.A.; Miranda, C. Optimization of edible coating composition to retard strawberry fruit senescence. *Postharvest Biol. Technol.* **2007**, *44*, 63–70. [CrossRef]
43. Harker, F.R.; Feng, J.; Johnston, J.W.; Gamble, J.; Alavi, M.; Hall, M.; Chheang, S.L. Influence of postharvest water loss on apple quality: The use of a sensory panel to verify destructive and non-destructive instrumental measurements of texture. *Postharvest Biol. Technol.* **2019**, *148*, 32–37. [CrossRef]
44. Tripathi, P.; Dubey, N.K.; Banerji, R.; Chansouria, J.P.N. Evaluation of some essential oils as botanical fungitoxicants in management of post-harvest rotting of citrus fruits. *World J. Microbiol. Biotechnol.* **2004**, *20*, 317–321. [CrossRef]
45. Basiak, E.; Linke, M.; Debeaufort, F.; Lenart, A.; Geyer, M. Dynamic behaviour of starch-based coatings on fruit surfaces. *Postharvest Biol. Technol.* **2019**, *147*, 166–173. [CrossRef]
46. Valencia-Chamorro, S.A.; Palou, L.; del Río, M.A.; Pérez-Gago, M.B. Antimicrobial edible films and coatings for fresh and minimally processed fruits and vegetables: A review. *Crit. Rev. Food Sci. Nutr.* **2011**, *51*, 872–900. [CrossRef] [PubMed]
47. Grande-Tovar, C.D.; Chaves-Lopez, C.; Serio, A.; Rossi, C.; Paparella, A. Chitosan coatings enriched with essential oils: Effects on fungi involve in fruit decay and mechanisms of action. *Trends Food Sci. Technol.* **2018**, *78*, 61–71. [CrossRef]
48. Campos-Requena, V.H.; Pérez, M.A.; Sanfuentes, E.A.; Figueroa, N.E.; Figueroa, C.R.; Rivas, B.L. Thermoplastic starch/clay nanocomposites loaded with essential oil constituents as packaging for strawberries—In vivo antimicrobial synergy over *Botrytis cinerea*. *Postharvest Biol. Technol.* **2017**, *129*, 29–36. [CrossRef]
49. Shao, X.; Cao, B.; Xu, F.; Xie, S.; Yu, D.; Wang, H. Effect of postharvest application of chitosan combined with clove oil against citrus green mold. *Postharvest Biol. Technol.* **2015**, *99*, 37–43. [CrossRef]

coatings

Article

Effect of Chitosan-Ascorbic Acid Coatings on the Refrigerated Storage Stability of Fresh-Cut Apples

Kübra Sultan Özdemir [1,2] and Vural Gökmen [1,*]

1 Food Quality and Safety (FoQuS) Research Group, Department of Food Engineering, Hacettepe University, Beytepe Campus, 06800 Ankara, Turkey
2 Department of Food Engineering, Faculty of Engineering and Architecture, Konya Food and Agriculture University, 42080 Meram, Konya, Turkey
* Correspondence: vgokmen@hacettepe.edu.tr; Tel.: +90-312-297-7108; Fax: +90-312-2992123

Received: 15 July 2019; Accepted: 7 August 2019; Published: 8 August 2019

Abstract: Using natural antimicrobial substances in edible films becomes crucial to extend the shelf-life of fresh-cut fruits due to the consumers' preferences. In this study chitosan and ascorbic acid based film was used to improve the shelf-life of fresh-cut apples. Fresh-cut apple cubes were dipped in water (control), ascorbic acid (1%) or mixtures of chitosan–ascorbic acid in different ratios (1%:1%, 2%:2% or 1%:5%) for 5 min. After draining, fresh-cut apples were packed in sterile polypropylene jars and stored at 5 °C for 14 days. The treatment with chitosan and ascorbic acid suppressed browning, retained flesh firmness and maintained phenolic compounds throughout the storage period. Moreover, the treatment with chitosan–ascorbic acid significantly retarded the microbial growth during storage. Those findings suggested that the best performance was acquired in 1% chitosan and 5% ascorbic acid coating. That coating could be practical and useful to prolonging the chemical and microbial shelf lives of fresh-cut apples during refrigerated storage.

Keywords: fresh-cut apples; refrigerated storage; shelf-life; chitosan; ascorbic acid; coating; Gompertz model

1. Introduction

Apple (*Malus domestica*) is an important—and one of the most widely consumed—fruits in the world, due to its health benefits. Apples contain significant amount of minerals, dietary fibers and polyphenols which are related to prevention of diseases, such as coronary diseases [1]. Polyphenols especially are related to the nutritional value, flavor and color of apple products. They can vary depending on the apple type [2]. The major polyphenols of apple can be listed as flavan-3-ols (catechin, epicatechin), flavanols (quercetin) dihydrochalcones, hydroxycinnamic acids (chlorogenic acid), and anthocyanins [3]. In recent years, the increasing demand of consumers for healthy fast and easily accessible foods, caused a strong increase in prepacked 'ready-to-eat' produce consumption [4]. Among those foods, fresh-cut apples have an important economic value in that market due to their low cost, convenience and popularity. However, microbial growth and browning reactions limit their shelf-life. Fresh cut apples are subject to enzymatic browning, due to the action of polyphenol oxidase on phenolic compounds. Browning causes undesired changes in appearance, flavor and nutrient composition [5]. The common approach to control/prevention of the browning in fresh-cut products is to use anti-browning agents [6]. Carboxylic acids, ascorbic acid derivatives, sulfur containing amino acids and phenolic acids can be used effectively as anti-browning agents to delay enzymatic browning in apples [7,8].

Many strategies have been developed for the preservation of, and increasing the quality of, fresh-cut and minimally processed fruits and vegetables. These include the use of modified atmosphere packaging [8], chemical treatments [9], thermal treatments [10], edible coatings [11–13] and hurdle technologies [12,14]. Among these methods, edible coatings have been recently gained more interest than others in food preservation. Edible films and coatings are natural polymers which control moisture transfer, gas exchange, respiration and oxidative reaction rates during handling, storage and transportation of the fresh produce [15,16]. Carbohydrates, proteins, gums, lipids or starch-based edible films can be used as a coating material for fresh-cut products.

Chitosan is a carbohydrate-based biopolymer. It is an amino polysaccharide composed of randomly distributed β-(1-4)-linked D-glucosamine and *N*-acetyl-D-glucosamine units [17–19]. Chitosan based edible films and coatings have great potential in the preservation of fruit and vegetables [20–22]. Chitosan and its derivatives have an ability to inhibit the growth of yeasts, molds and bacteria [23–25]. However, the application of chitosan is limited due to its insolubility at neutral pH [26]. To overcome this disadvantage and to improve solubility of chitosan, different acids were used to decrease pH values below pH 6 to improve chitosan solubility [16,21,27–29].

Qi et al. 2011 [30] studied the effect of chitosan, ascorbic acid and calcium chloride on fresh-cut fuji apples. They used citric acid to dissolve chitosan and prepared film solutions by adding calcium chloride and ascorbic acid. They have found out that coatings effectively retarded enzymatic browning during storage at 5 °C for 8 days, and apple slices with chitosan-coating maintained firmness with a little loss during storage. In another study, calcium chloride-chitosan coatings and chitosan-ascorbic acid coatings during storage at room temperature for 10 h were investigated [31]. Calcium treatments (calcium chloride or lactate) are broadly studied in preserving the quality of fresh-cut fruit and vegatables. There are many reports about the positive effects of calcium on fresh-cut texture and browning [32]. This study aims to use a binary aqueous mixture of chitosan and ascorbic acid as a coating to improve the refrigerated storage stability of fresh-cut apple cubes. Contrary to other studies mentioned above, we used solely ascorbic acid instead of other organic acids (i.e., acetic acid) to dissolve chitosan in water. Lowering pH by means of adding ascorbic acid helped chitosan molecules to dissolve readily in water. In addition, ascorbic acid acted as anti-browning agent in the coating mixture. In this context, different combinations of chitosan and ascorbic acid were prepared and used to coat fresh-cut apples. The effects of different coatings were evaulated by monitoring the microbial growth, browning, polyphenol composition and firmness of fresh-cut apples during 14 days of storage at 5 °C. In addition, the kinetics of the growths of yeasts and molds, psychrophilic and mesophilic bacteria in fresh-cut apple samples with different coatings during storage were analyzed by using the Gompertz model.

2. Materials and Methods

2.1. Chemicals and Consumables

Ascorbic acid, chitosan (from crab shells, degree of deacetylation of 75–85% and medium molecular weight), ethanol (HPLC grade), acetonitrile (HPLC grade), chlorogenic acid (5-Caffeoylquinic acid), (+)-catechin, and (−) epicatechin were purchased from Sigma-Aldrich (Steinheim, Germany). Plate count agar (PCA), potato dextrose agar (PDA) and formic acid (98%) were obtained from Merck Co. (Darmstadt, Germany).

Ultra-pure water was used throughout the phenolic composition analysis (Milli Q-System, Millipore, Milford, MA, USA). Syringe filters (nylon, 0.45 μm) were supplied from Waters (Millford, MA). A HICHROM 5C 18 (250 × 4.6 mm, Hichrom, Reading, UK) column was used for analysis of phenolic compounds.

2.2. Preparation of Fresh-Cut Apples and Dipping Solutions

Apples (Granny Smith) were obtained from a local market at commercial maturity. Uniformly sized apples were selected and any bruised or diseased fruits were discarded. The selected apples were rinsed gently with tap water and drained at room temperature. Then apples were cut into about 1 cm^3 cubes with a stainless-steel knife.

Fresh-cut apples were dipped into water (control), aqueous ascorbic acid solution (1%ASC) or the mixtures of chitosan–ascorbic acid. The concentrations of chitosan and ascorbic acid were varied from 1 to 2% and from 1 to 5%, respectively, in three mixtures. In the first mixture, 1 g of ascorbic acid was dissolved in water and 1 g of chitosan was added to the solution (1%CH–1%ASC). In the second mixture, 5 g of ascorbic acid was dissolved in water and 1 g of chitosan was added to the solution (1%CH–5%ASC). In the third mixture, 2 g of ascorbic acid was dissolved in water and 2 g of chitosan was added to the solution (2%CH–2%ASC). Ultrasonification was applied after magnetic stirring to achieve complete dissolution of CH. pH values of solutions were measured by using a pH-meter at room temperature.

2.3. Treatments of Fresh-Cut Apples

Fresh-cut apple cubes were grouped into five and the treatments were applied. Each group (50 g) were dipped in 100 mL of solution or mixture for each time. After the application, the coated apples were left to surface dry (at about 25 °C and with a relative humidity of 30% during 1 h) until excess coating solution was gone. A total of 27 packages (six for microbiological, three for weight loss, and 18 for physical and chemical analyses) were prepared for each treatment. Apple samples of 10 g were placed into sterile polypropylene jars (30 mL). Biaxially oriented polypropylene (BOPP) film (thickness of 30 μm and O_2 permeability of 1600 $cm^3.m^{-2}$ per day) were used for covering the packages. All samples were stored at 5 ± 1 °C at 90–95% relative humidity for 14 days and samples were taken after 0, 3, 7 and 14 days of storage. Concentrations of phenolic compounds were determined at 0, 6, 12 and 18 days of storage.

2.4. Microbiological Analyses and Kinetic Modelling of Microbial Growth

The growth of microbial population in coated fresh-cut apples throughout storage was evaluated by counting total yeasts and molds, total aerobic mesophilic bacteria and psychrophilic bacteria. Apple samples of 10 g were taken aseptically from each jar and transferred into sterile plastic bags. Samples were diluted with 90 mL of saline water and homogenized for 2 min in a stomacher blender.

Serial dilutions were made and then poured onto the PCA for total aerobic mesophilic counts (TAC) and psychrophilic counts. TAC plates were incubated at 35 °C for 48 h, and to determine psychrophilic counts, plates were incubated at 5 °C for 10–15 days. Yeast and mold counts were performed on PDA by using the spread plate method by incubating at 25 °C for 5–7 days. Colonies were counted and the results were given as $\log_{10}$ $CFU.g^{-1}$ of apples. Two replicate counts were performed for each sample. Moreover, the Gompertz equation modified by Zwietering *et al.*(1990) [33], was used to describe the microbial data from each application and given in Equation (1):

$$\log_{10}\left(\frac{N}{N_0}\right) = A\exp\left(-\exp\left(\frac{\mu_m e}{A}(\lambda - t) + 1\right)\right) \tag{1}$$

where N is the number of microorganisms at time t (day), N_0 is initial number of microorganisms, μ_m is the maximum growth rate, λ is the lag time (day), A is the maximum population density and t is the time (day). The experimental data were fitted by using Matlab version 9.2.

2.5. Measurement of Weight Loss

Fresh-cut apples were weighed at the end of days 0, 3, 7 and 14. Measurements were replicated three times. Considering of the initial and final weights of fresh-cut apples, weight loss was determined as expressed in Equation (2):

$$\textit{Weight loss } (\%) = \frac{(m_t - m_0)}{m_0} \times 100 \tag{2}$$

where m_0 is the initial weight of fresh-cut apples and m_t is the weight of sample at time t.

2.6. Measurement of Firmness

Firmness of fresh-cut apples was determined by using TA Plus texture analyzer (LLYOD Instruments, AMETEK Co. Ltd., Shanghai, China) with a 5 mm diameter cylindrical probe. The required force (N) to puncture the samples was recorded as firmness. Measurements were replicated six times.

2.7. Measurement of Color

For the effect of coatings on color, browning index (BI) values were calculated by using computer-vision based image analyses. Apple cubes were photographed from 25 cm height above the sample by a digital camera (Canon, Japan) in a box. Two light sources (Ultrabright, 6500 K, 25 W) were placed at a 45° angle to the sample. The percentage of dark areas was calculated by segmentation method given by Gökmen and Mogol (2010) and data were given as browning index [34].

2.8. Analysis of Phenolic Compounds

One gram of apple sample was homogenized in 5 mL of the mixture of ethanol and water (50:50, *v*/*v*) containing 500 mg/L ascorbic acid. After centrifugation (6080 *g* for 5 min) supernatants were separated in a test tube. Polyphenol analyses were performed on an Agilent 1200 LC system equipped with an Agilent 6130 MS detector. Chromatographic separation was performed on a HICHROM 5 C18 column at 30 °C according to the method given by Ozdemir et al. [35]. Solution A was 1.0% formic acid in water and solution was B 1.0% formic acid in acetonitrile. The gradient profile was as following: 0–8 min linear gradient elution from 10 to 30% of B; 10–12 min linear gradient elution from 30 to 10% B; and 12–20 min; isocratic elution of 10% B. For analysis, 10 μL of sample was injected and the flow rate was 0.7 mL/min. An electrospray ionization source was used and the phenolic compounds were analyzed in negative and positive modes. The following MS conditions were adjusted: Positive scan mode of capillary voltage 4.0 kV; negative scan mode of capillary voltage 3.5 kV; drying gas (N_2) flow of 13 L/min at 325 °C; nebulizer pressure of 40 psi; negative and positive ion scanning modes from 50 to 1000 m/z. Phenolic compounds were identified by comparing the retention times of unknown peaks with the retention times of standard compounds.

2.9. Statistical Analysis

The results were given as mean ± standard deviations. Differences were determined by analysis of variance (ANOVA) followed by Duncan's multiple range test ($p < 0.05$). Independent sample t testing was carried out to compared the means of two independent samples. For the statistical analyses, SPSS 18.0 version (SPSS Inc., Chicago, IL, USA) was used.

3. Results and Discussion

3.1. Weight Loss

The weight of coated fresh-cut apples was monitored to determine the efficiency of these coatings during storage time in terms of keeping the moisture. It was known that weight loss in fresh-cut fruits is related to the loss of water caused by transpiration and loss of carbon reserves due to the respiration processes [36]. The weight losses of coated samples with 1%CH–1%ASC and 1%CH–5%ASC were significantly lower than those of the control and 1% ASC treated apples at the end of storage ($p < 0.05$) (Table 1). The highest weight loss was determined in 1% ASC coated apples. Furthermore, 1%CH–1%ASC coating showed the best performance on inhibition of water loss by showing the lowest weight loss at the end of storage.

Table 1. Changes in the weight loss (%) values of fresh cut apples during storage at 5 °C *.

Treatment	Storage Time (day)		
	3	7	14
Control	0.47 ± 0.08 aA	0.55 ± 0.23 aAB	3.45 ± 0.40 bC
1%ASC	0.45 ± 0.01 aA	0.31 ± 0.02 aA	5.38 ± 0.39 bD
1%CH–1%ASC	1.04 ± 0.12 aB	1.00 ± 0.24 aB	1.31 ± 0.26 aA
1%CH–5%ASC	0.94 ± 0.26 aB	0.92 ± 0.31 aB	2.11 ± 0.72 aAB
2%CH–2%ASC	0.87 ± 0.04 aB	1.65 ± 0.18 bC	2.63 ± 0.33 cBC

* Values were presented in mean ± standard deviation (n = 3). Different lowercase and uppercase letters indicate significance at α = 0.05 in the same row and column, respectively.

This result was related to the formation of a layer of chitosan around the surface of fruit. This layer slowed down the respiration rate of the fruit and retarded the loss of water. Our results supported by other studies revealed that chitosan coating reduced weight loss in fruits and vegetables, such as mango fruit [28], bell pepper, cucumber [37] and litchi [38].

3.2. Firmness

In the present study, the firmness of the control, %1ASC and 1%CH–1%ASC treated apples did not significantly decrease in the 7 days of storage but after 14 days, data could not be obtained due to the substantial softening of apples (Table 2). That softening was related with the action of pectin enzymes in apple samples.

Table 2. Changes in the firmness (N) values of fresh cut apples during storage at 5 °C *.

Treatment	Storage Time (Day)			
	1	3	7	14
Control	20.2 ± 2.2 bA	17.6 ± 0.3 aA	18.6 ± 1.8 abA	–
1%ASC	20.9 ± 1.7 aA	19.6 ± 2.6 aA	20.8 ± 1.8 aA	–
1%CH–1%ASC	20.6 ± 1.2 aA	19.0 ± 2.8 aA	19.6 ± 2.2 aA	–
1%CH–5%ASC	19.6 ± 2.1 aA	20.1 ± 2.3 aA	19.3 ± 2.3 aA	18.1 ± 3.5 aA
2%CH–2%ASC	19.1 ± 1.7 aA	19.4 ± 1.9 aA	19.3 ± 2.4 aA	18.4 ± 4.6 aA

* Values were presented in mean ± standard deviation (n = 6). Different lowercase and uppercase letters indicate significance at α = 0.05 in the same row and column, respectively.

Polygalacturonase enzymes especially, cause changes in texture and loss of firmness of the fruit [39,40]. However, firmness of 1%CH–5%ASC and 2%CH–2%ASC treated apples remained stable during 14 days of storage. This can be explained with increment in chitosan and ascorbic acid concentrations. 1% ASC and 1%CH–1%ASC treated apples lost their textural integrity faster than apples coated with 1%CH–5%ASC and 2%CH–2%ASC. Those results are in line with the results obtained by other researchers, which underline the beneficial effects of chitosan toward fruit firmness maintenance [30,40].

3.3. Color

Data of browning index (BI) values calculated by using computer-vision based image analyses are given in Figure 1. BI was increased with time during storage at 5 °C for all applications ($p < 0.05$).

Figure 2 shows the digital color images of fresh-cut apple samples with different treatments. BI increased in control samples immediately after cutting apples, but chitosan–ascorbic coating retarded the browning of the fruit. The browning index increased up to 64% in control samples and 54% in 1%ASC treated samples while it was lower in 1%CH–5%ASC and 2%CH–2%ASC coated apples. %1 ascorbic acid had also a positive effect on the browning ratio but in combination with chitosan, it was more effective over storage time. Although browning occurred in chitosan–ascorbic acid combinations at the end of storage, 1%CH–5%ASC and 2%CH–2%ASC coatings retarded the browning in the first three days.

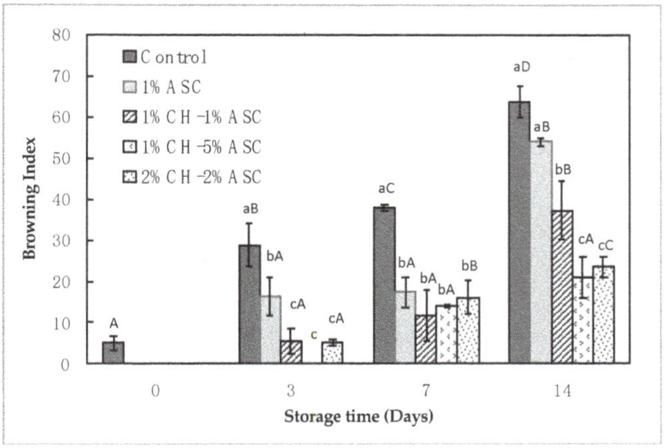

Figure 1. Changes in the browning index (%) values of fresh cut apple samples during storage at 5 °C. Values are means of two replications. Vertical bars represent standard deviation. Data in the same day with each treatment having different superscript lowercase letters are significantly different ($p < 0.05$). Data on each day of storage in the same treatment having different superscript uppercase letters are significantly different ($p < 0.05$).

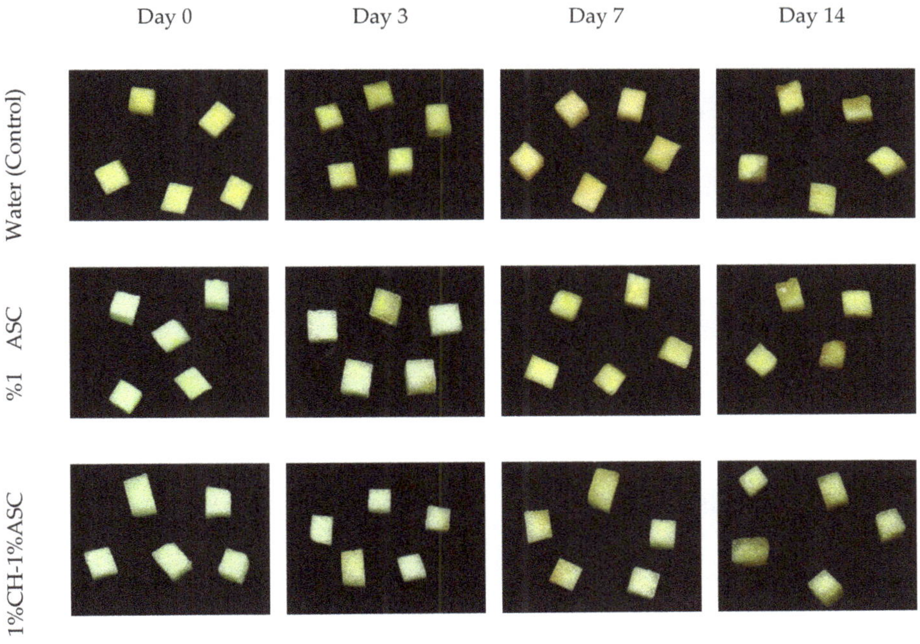

Figure 2. *Cont.*

Figure 2. Images of fresh-cut apples after coating and storing at 5 °C for days 0, 3, 7 and 14.

Chitosan–ascorbic acid coatings inhibited the browning of apple pieces, and browning inhibition was enhanced by increasing the ascorbic acid and chitosan ratio in the solution. These results were also supported by the studies on the beneficial side of chitosan on browning [41–43]. In addition, anti-browning agents such as ascorbic acid and citric acid suppress the polyphenol oxidase activity working as reducing agents [44].

3.4. Phenolic Compounds

The most important flavanol compounds in apples, catechin, epicatechin and chlorogenic acid (5-Caffeoylquinic acid), were determined, and concentrations (C/C_0) of those polyphenols were presented in Table 3. Those compounds are strongly related with browning, because they are well-known substrates to the polyphenol oxidase enzyme [45]. The concentrations of epicatechin and catechin in control samples decreased at a ratio of 70% and 90%, respectively, at the end of storage. However, initial concentrations of catechin and epicatechin could be retained 89% and 71%, respectively, in the samples coated with the mixture of %1CH–5% ASC (Table 3).

Table 3. Changes in the dimensionless concentrations (C/C_o) of certain phenolic compounds in fresh cut apples during storage at 5 °C *.

Analysis	Day	Control	1%CH–1%ASC	1%CH–5%ASC
Catechin	0	1.00 ± 0.00 ^{aA}	1.00 ± 0.00 ^{aA}	1.00 ± 0.00 ^{abA}
	6	0.17 ± 0.02 ^{bA}	0.80 ± 0.12 ^{bB}	1.17 ± 0.09 ^{aC}
	12	0.13 ± 0.01 ^{bcA}	0.21 ± 0.02 ^{cA}	0.95 ± 0.09 ^{bB}
	18	0.10 ± 0.02 ^{cA}	0.20 ± 0.00 ^{cA}	0.89 ± 0.07 ^{bB}
Epicatechin	0	1.00 ± 0.00 ^{aA}	1.00 ± 0.0 ^{aA}	1.00 ± 0.00 ^{aA}
	6	0.35 ± 0.07 ^{bcA}	0.89 ± 0.09 ^{aB}	0.75 ± 0.11 ^{abB}
	12	0.46 ± 0.00 ^{bA}	0.42 ± 0.01 ^{bA}	0.55 ± 0.09 ^{bA}
	18	0.30 ± 0.00 ^{cA}	0.28 ± 0.03 ^{cA}	0.71 ± 0.19 ^{abB}
Chlorogenic acid	0	1.00 ± 0.00 ^{aA}	1.00 ± 0.00 ^{aA}	1.00 ± 0.00 ^{aA}
	6	0.93 ± 0.00 ^{aA}	1.05 ± 0.11 ^{aA}	1.06 ± 0.02 ^{aA}
	12	0.99 ± 0.08 ^{aA}	1.30 ± 0.04 ^{aAB}	0.82 ± 0.01 ^{bBC}
	18	1.06 ± 0.07 ^{aA}	1.12 ± 0.04 ^{aA}	0.91 ± 0.13 ^{abA}

* Values were presented in mean ± standard deviation (n = 2). Different lowercase and uppercase letters indicate significance at α = 0.05 in the same column and row, respectively.

Chlorogenic acid was not significantly affected ($p > 0.05$) by the storage period for all applications (Table 3). According to the results the best application to retain polyphenol content in apples was 1% chitosan and 5% ascorbic acid formulation. Ascorbic acid treatment combined with chitosan preserved polyphenol content to maintain the same levels as the first day. Moreover, increasing the ascorbic acid proportion from 1% to 5% in the coating solution was more effective for maintaining phenolic compounds. Similarly, ascorbic acid with carboxymethyl cellulose coating maintained phenolic compounds in fresh-cut apple slices [7]. It has been reported that ascorbic acid ensures protection by its oxygen scavenger property and avoiding polyphenol oxidase-catalyzed reactions [46].

3.5. Microbialanalyses and Shelf Life Modelling

In the first day of storage, in all samples, total aerobic mesophilic bacteria, total yeasts and molds were found to be below the detectable amount. In the control sample and 1% ASC treated sample, the number of total yeasts and molds was determined as 6.5 ± 0.4 log cfu/g and 6.9 ± 0.2 log cfu/g at the end of storage, respectively (Table 4). To evaluate the results on Table 4, pH values of coating solutions were also measured and determined as: Control solution: 6.80; 1%CH–1%ASC: 4.42; 2%CH–2%ASC: 4.45; 1%CH–5%ASC: 3.31; 1% ascorbic acid: 2.92; all at room temperature. Cadogan et al. 2014 investigated the effect of solvents on antibacterial activity of chitosan membranes. They showed that the chitosan dissolved in ascorbic acid had higher antimicrobial activity than the chitosan dissolved in citric and maleic acid. This was related with the crosslinking effect of ascorbic acid. Polymer molecules, i.e., chitosan, could have easily penetrated into bacterial cell walls and hence changed the internal pH of bacteria [47]. In our study, the ratio of chitosan was the same in 1% CH– 5%ASC and 1%CH–1% ASC coatings. Their pH levels were 3.31 and 4.45, respectively. Treatments of 1%CH–5%ASC showed no yeast and mold growth in day 14 while 3.2 log CFU/g was observed in 1%CH–1%ASC. This could be related with pH and crosslinking properties of ascorbic acid. However, the 1% CH–1% ASC and 2%CH–2%ASC coatings' pH levels were 4.42 and 4.45, yet the antimicrobial activity was higher in 2%CH–2%ASC. This could be derived from the higher amount of chitosan.

Table 4. Changes in the counts of yeasts and molds, aerobic mesophilic bacteria and psychrophilic bacteria ($\log_{10}$ CFU/g) in fresh-cut apples during storage at 5 °C *.

Treatment	Day 0	Day 7	Day 14
	Yeasts and molds		
Control	UDL *	$4.0 \pm 0.2^{**aA}$	6.5 ± 0.4^{aB}
1% ASC	UDL	5.0 ± 0.0^{aA}	6.9 ± 0.2^{aB}
1% CH–1% ASC	UDL	2.5 ± 0.3^{bA}	3.2 ± 0.2^{bA}
2% CH–2% ASC	UDL	UDL	UDL
1% CH–5% ASC	UDL	UDL	UDL
	Aerobic mesophilic bacteria		
Control	UDL	1.0 ± 0.0^{A}	3.1 ± 0.1^{B}
1% ASC	UDL	UDL	UDL
1% CH–1% ASC	UDL	UDL	UDL
2% CH–2% ASC	UDL	UDL	UDL
1% CH–5% ASC	UDL	UDL	UDL
	Psychrophilic bacteria		
Control	1.0 ± 0.0^{aA}	4.0 ± 0.4^{aB}	5.2 ± 0.1^{aC}
1% ASC	1.1 ± 0.3^{aA}	4.0 ± 0.4^{aB}	5.1 ± 0.3^{aC}
1% CH–1% ASC	1.0 ± 0.0^{aA}	1.9 ± 0.3^{bB}	UDL
2% CH–2% ASC	1.2 ± 0.2^{a}	UDL	UDL
1% CH–5% ASC	UDL	UDL	UDL

* UDL:Under Detection Limit < 1 $\log_{10}$ CFU/g ** Values were presented in mean ± standard deviation (n = 4). Different lowercase and uppercase letters indicate significance at $\alpha = 0.05$ in the same column and row, respectively.

The quantity of aerobic mesophilic bacteria at the end of the storage was 3.1 log cfu/g in control samples, while it was found to be below the detectable amount in coated samples. In that situation, it can be said that chitosan film coatings with ascorbic acid reduced or inhibited the total aerobic mesophilic bacteria and total yeasts and molds. The number of total psychrophilic aerobic bacteria were increased in the control and 1% ASC treated apple samples up to 5.2 log CFU/g at the end of storage, while it was determined below the detectable count in chitosan and ascorbic acid coated samples. Also, many researchers determined that chitosan coatings reduce microbial counts in fresh-cut produce [18,48–50].

The microbiological data were described by using the Gompertz equation. The three interpretable parameters defined by the model (maximum growth rate, lag time and maximum population density) are given in Table 5. The model described the growth of yeast and mold counts, as well as those of mesophilic aerobic counts with high determination coefficients (R^2). The maximum growth rate and maximum population density were highly influenced by the different treatments. The maximum growth rate for yeast and molds were determined as 1.16 and 1.38 Δ $\log_{10}(CFU.g^{-1})$/day for control and 1% ASC treated samples, respectively, while it was 0.49 for 1%CH-1%ASC coated apples. Those results confirmed the microbiological counts and it was seen that microbial growth was faster in control and 1%ASC coated samples than other applications. Eventually, maximum growth (A) was higher for yeast and molds in control and 1%ASC treated apples than that found in 1%CH–1%ASC treated apples. That was also related with antimicrobial activity of chitosan.

Table 5. Gompertz model parameters for the growth of yeasts and molds, psychrophilic and mesophilic bacteria in fresh-cut apples during storage at 5 °C *.

Population	Treatments	Gompertz Model Parameters			
		A (log cfu/g)	μ_{max} (Δ log [cfu/g]/day)	λ (day)	R^2
Yeasts and molds	Control	6.63 ± 0.46	1.16 ± 0.09	3.52 ± 0.05	1.00
	1% ASC	6.91 ± 0.20	1.38 ± 0.01	3.08 ± 0.09	1.00
	1%CH-1%ASC	2.99 ± 0.42	0.49 ± 0.07	2.86 ± 0.70	1.00
	2%CH-2%ASC	–	–	–	–
	1%CH-5%ASC	–	–	–	–
Mesophilic aerobic bacteria	Control	3.32 ± 0.10	0.46 ± 0.00	4.79 ± 0.02	1.00
	1% ASC	–	–	–	–
	1%CH-1%ASC	–	–	–	–
	2%CH-2%ASC	–	–	–	–
	1%CH-5%ASC	–	–	–	–
Psychrophilic bacteria	Control	4.21 ± 0.11	0.94 ± 0.10	3.50 ± 0.21	1.00
	1% ASC	3.96 ± 0.04	0.94 ± 0.20	3.46 ± 0.63	1.00
	1%CH-1%ASC	–	–	–	–
	2%CH-2%ASC	–	–	–	–
	1%CH-5%ASC	–	–	–	–

* Gompertz model parameters; A: maximum population density; μ_{max}: maximum growth rate; λ: lag time; R^2: coefficient of determination. Significance level at $p < 0.05$. Values are given in mean ± standard deviation of four means. (–) Data not obtained due to the microbial growth under detection limit.

4. Conclusions

The results revealed that coating with chitosan and ascorbic acid was an effective solution for inhibiting the growth of microorganisms, retarding enzymatic browning reactions and reducing the weight loss of fresh-cut apples during refrigerated storage. Increasing the ratio of ascorbic acid in the coatings decreased the browning ratio on fresh-cut apples. From a practical point of view, 1% chitosan and 5% ascorbic acid coating successfully extended the cold storage period for 7 days in fresh-cut apples by maintaining the browning level, preserving microbial and chemical quality. The results suggest that coating with chitosan and ascorbic acid can be a viable approach to improve the shelf life of fresh-cut fruits sensitive to microbial and chemical deterioration under refrigerated conditions. Additionally, such an approach could be used in combination with modified atmosphere packaging to further prolong the shelf life.

Patents

WO2015142303A1: A solution for extending shelf life of ready-to-eat fresh fruits and/or vegetables and an application method thereof

Author Contributions: Writing—original draft preparation, K.S.Ö.; investigation K.S.Ö, and V.G.; methodology, K.S.Ö, and V.G.; writing—review and editing, V.G.

Funding: This research received no external funding.

Conflicts of Interest: The authors declare no conflict of interest.

References

1. Marszałek, K.; Woźniak, Ł.; Barba, F.J.; Skąpska, S.; Lorenzo, J.M.; Zambon, A.; Spilimbergo, S. Enzymatic, physicochemical, nutritional and phytochemical profile changes of apple (Golden Delicious L.) juice under supercritical carbon dioxide and long-term cold storage. *Food Chem.* **2018**, *268*, 279–286. [CrossRef] [PubMed]
2. Li, X.; Wu, X.; Bi, J.; Liu, X.; Li, X.; Guo, C. Polyphenols accumulation effects on surface color variation in apple slices hot air drying process. *LWT* **2019**, *108*, 421–428. [CrossRef]
3. Guo, J.; Yue, T.L.; Yuan, Y.H.; Wang, Y.T. Chemometric Classification of Apple Juices According to Variety and Geographical Origin Based on Polyphenolic Profiles. *J. Agric. Food Chem.* **2013**, *61*, 6949–6963. [CrossRef] [PubMed]
4. Kabelitz, T.; Hassenberg, K. Control of apple surface microflora for fresh-cut produce by post-harvest hot-water treatment. *LWT* **2018**, *98*, 492–499. [CrossRef]
5. Luo, Y.; Barbosa Canovas, G.V. Enzymatic browning and its inhibition in new apple cultivars slices using 4-hexylresorcinol in combination with ascorbic acid. *Food Sci. Technol. Int.* **1997**, *3*, 195–201. [CrossRef]
6. Mcevily, A.J.; Iyengar, R.; Otwell, W.S. Inhibition of Enzymatic Browning in Foods and Beverages. *Crit. Rev. Food Sci.* **1992**, *32*, 253–273. [CrossRef]
7. Saba, M.K.; Sogvar, O.B. Combination of carboxymethyl cellulose-based coatings with calcium and ascorbic acid impacts in browning and quality of fresh-cut apples. *LWT Food Sci. Technol.* **2016**, *66*, 165–171. [CrossRef]
8. Putnik, P.; Kovacevic, D.B.; Herceg, K.; Pavkov, I.; Zoric, Z.; Levaj, B. Effects of modified atmosphere, anti-browning treatments and ultrasound on the polyphenolic stability, antioxidant capacity and microbial growth in fresh-cut apples. *J. Food Process. Eng.* **2017**, *40*, e12359. [CrossRef]
9. Oms-Oliu, G.; Aguiló-Aguayo, I.; Martín-Belloso, O. Inhibition of Browning on Fresh-cut Pear Wedges by Natural Compounds. *J. Food Sci.* **2006**, *71*, S216–S224. [CrossRef]
10. Terefe, N.S.; Tepper, P.; Ullman, A.; Knoerzer, K.; Juliano, P. High pressure thermal processing of pears: Effect on endogenous enzyme activity and related quality attributes. *Innov. Food Sci. Emerg.* **2016**, *33*, 56–66. [CrossRef]
11. Alves, M.M.; Goncalves, M.P.; Rocha, C.M.R. Effect of ferulic acid on the performance of soy protein isolate-based edible coatings applied to fresh-cut apples. *LWT Food Sci. Technol.* **2017**, *80*, 409–415. [CrossRef]
12. Sanchis, E.; Ghidelli, C.; Sheth, C.C.; Mateos, M.; Palou, L.; Perez-Gago, M.B. Integration of antimicrobial pectin-based edible coating and active modified atmosphere packaging to preserve the quality and microbial safety of fresh-cut persimmon (Diospyros kaki Thunb. cv. Rojo Brillante). *J. Sci. Food Agric.* **2017**, *97*, 252–260. [CrossRef] [PubMed]
13. Marquez, G.R.; Di Pierro, P.; Mariniello, L.; Esposito, M.; Giosafatto, C.V.L.; Porta, R. Fresh-cut fruit and vegetable coatings by transglutaminase-crosslinked whey protein/pectin edible films. *LWT Food Sci. Technol.* **2017**, *75*, 124–130. [CrossRef]
14. Moreira, M.R.; Alvarez, M.V.; Martin-Belloso, O.; Soliva-Fortuny, R. Effects of pulsed light treatments and pectin edible coatings on the quality of fresh-cut apples: A hurdle technology approach. *J. Sci. Food Agric.* **2017**, *97*, 261–268. [CrossRef] [PubMed]
15. Rojas-Grau, M.A.; Sobrino-Lopez, A.; Tapia, M.S.; Martin-Belloso, O. Browning inhibition in fresh-cut 'fuji' apple slices by natural antibrowning agents. *J. Food Sci.* **2006**, *71*, S59–S65. [CrossRef]
16. Park, H.J. Development of advanced edible coatings for fruits. *Trends Food Sci. Technol.* **1999**, *10*, 254–260. [CrossRef]
17. Xia, W.; Liu, P.; Zhang, J.; Chen, J. Biological activities of chitosan and chitooligosaccharides. *Food Hydrocoll.* **2011**, *25*, 170–179. [CrossRef]
18. Devlieghere, F.; Vermeulen, A.; Debevere, J. Chitosan: Antimicrobial activity, interactions with food components and applicability as a coating on fruit and vegetables. *Food Microbiol.* **2004**, *21*, 703–714. [CrossRef]

19. Dutta, P.K.; Tripathi, S.; Mehrotra, G.K.; Dutta, J. Perspectives for chitosan based antimicrobial films in food applications. *Food Chem.* **2009**, *114*, 1173–1182. [CrossRef]
20. Kerch, G. Chitosan films and coatings prevent losses of fresh fruit nutritional quality: A review. *Trends Food Sci. Technol.* **2015**, *46*, 159–166. [CrossRef]
21. Wang, S.Y.; Gao, H. Effect of chitosan-based edible coating on antioxidants, antioxidant enzyme system, and postharvest fruit quality of strawberries (Fragaria x aranassa Duch.). *LWT Food Sci. Technol.* **2013**, *52*, 71–79. [CrossRef]
22. Ma, L.; Zhang, M.; Bhandari, B.; Gao, Z.X. Recent developments in novel shelf life extension technologies of fresh-cut fruits and vegetables. *Trends Food Sci. Technol.* **2017**, *64*, 23–38. [CrossRef]
23. Coma, V.; Martial-Gros, A.; Garreau, S.; Copinet, A.; Salin, F.; Deschamps, A. Edible Antimicrobial Films Based on Chitosan Matrix. *J. Food Sci.* **2002**, *67*, 1162–1169. [CrossRef]
24. Shahidi, F.; Arachchi, J.K.V.; Jeon, Y.-J. Food applications of chitin and chitosans. *Trends Food Sci. Technol.* **1999**, *10*, 37–51. [CrossRef]
25. Vásconez, M.B.; Flores, S.K.; Campos, C.A.; Alvarado, J.; Gerschenson, L.N. Antimicrobial activity and physical properties of chitosan–tapioca starch based edible films and coatings. *Food Res. Int.* **2009**, *42*, 762–769. [CrossRef]
26. Ge, H.-C.; Luo, D.-K. Preparation of carboxymethyl chitosan in aqueous solution under microwave irradiation. *Carbohydr. Res.* **2005**, *340*, 1351–1356. [CrossRef] [PubMed]
27. Chien, P.-J.; Sheu, F.; Yang, F.-H. Effects of edible chitosan coating on quality and shelf life of sliced mango fruit. *J. Food Eng.* **2007**, *78*, 225–229. [CrossRef]
28. Jongsri, P.; Wangsomboondee, T.; Rojsitthisak, P.; Seraypheap, K. Effect of molecular weights of chitosan coating on postharvest quality and physicochemical characteristics of mango fruit. *LWT Food Sci. Technol.* **2016**, *73*, 28–36. [CrossRef]
29. Tezotto-Uliana, J.V.; Fargoni, G.P.; Geerdink, G.M.; Kluge, R.A. Chitosan applications pre- or postharvest prolong raspberry shelf-life quality. *Postharvest Biol. Technol.* **2014**, *91*, 72–77. [CrossRef]
30. Qi, H.P.; Hu, W.Z.; Jiang, A.L.; Tian, M.X.; Li, Y.Q. Extending shelf-life of Fresh-cut 'Fuji' apples with chitosan-coatings. *Innov. Food Sci. Emerg.* **2011**, *12*, 62–66. [CrossRef]
31. Liu, X.; Ren, J.; Zhu, Y.; Han, W.; Xuan, H.; Ge, L. The preservation effect of ascorbic acid and calcium chloride modified chitosan coating on fresh-cut apples at room temperature. *Coll. Surf. A Physicochem. Eng. Asp.* **2016**, *502*, 102–106. [CrossRef]
32. Ngamchuachit, P.; Sivertsen, H.K.; Mitcham, E.J.; Barrett, D.M. Effectiveness of calcium chloride and calcium lactate on maintenance of textural and sensory qualities of fresh-cut mangos. *J. Food Sci.* **2014**, *79*, C786–C794. [CrossRef]
33. Zwietering, M.H.; Jongenburger, I.; Rombouts, F.M.; van't Riet, K. Modeling of the bacterial growth curve. *Appl. Environ. Microbiol.* **1990**, *56*, 1875–1881. [PubMed]
34. Gökmen, V.; Mogol, B.A. Computer vision-based image analysis for rapid detection of acrylamide in heated foods. *Qual. Assur. Saf. Crop. Foods* **2010**, *2*, 203–207. [CrossRef]
35. Özdemir, K.S.; Yılmaz, C.; Durmaz, G.; Gökmen, V. Hazelnut skin powder: A new brown colored functional ingredient. *Food Res. Int.* **2014**, *65*, 291–297. [CrossRef]
36. Zhu, X.; Wang, Q.M.; Cao, J.K.; Jiang, W.B. Effects of chitosan coating on postharvest quality of mango (Mangifera indica L. cv. Tainong) fruits. *J. Food Process. Preserv.* **2008**, *32*, 770–784. [CrossRef]
37. Elghaouth, A.; Arul, J.; Ponnampalam, R.; Boulet, M. Use of Chitosan Coating to Reduce Water-Loss and Maintain Quality of Cucumber and Bell Pepper Fruits. *J. Food Process. Preserv.* **1991**, *15*, 359–368.
38. Dong, H.Q.; Cheng, L.Y.; Tan, J.H.; Zheng, K.W.; Jiang, Y.M. Effects of chitosan coating on quality and shelf life of peeled litchi fruit. *J. Food Eng.* **2004**, *64*, 355–358. [CrossRef]
39. Raybaudi-Massilia, R.M.; Mosqueda-Melgar, J.; Sobrino-López, A.; Soliva-Fortuny, R.; Martín-Belloso, O. Shelf-life extension of fresh-cut "Fuji" apples at different ripeness stages using natural substances. *Postharvest Biol. Technol.* **2007**, *45*, 265–275. [CrossRef]
40. Hong, K.Q.; Xie, J.H.; Zhang, L.B.; Sun, D.Q.; Gong, D.Q. Effects of chitosan coating on postharvest life and quality of guava (Psidium guajava L.) fruit during cold storage. *Sci. Hortic.* **2012**, *144*, 172–178. [CrossRef]
41. Hosseini, M.S.; Zahedi, S.M.; Abadía, J.; Karimi, M. Effects of postharvest treatments with chitosan and putrescine to maintain quality and extend shelf-life of two banana cultivars. *Food Sci. Nutr.* **2018**, *6*, 1328–1337. [CrossRef] [PubMed]

42. Gao, P.; Zhu, Z.; Zhang, P. Effects of chitosan–glucose complex coating on postharvest quality and shelf life of table grapes. *Carbohydr. Polym.* **2013**, *95*, 371–378. [CrossRef] [PubMed]
43. Carvalho, R.L.; Cabral, M.F.; Germano, T.A.; de Carvalho, W.M.; Brasil, I.M.; Gallão, M.I.; Moura, C.F.H.; Lopes, M.M.A.; de Miranda, M.R.A. Chitosan coating with trans-cinnamaldehyde improves structural integrity and antioxidant metabolism of fresh-cut melon. *Postharvest Biol. Technol.* **2016**, *113*, 29–39. [CrossRef]
44. Ali, H.M.; El-Gizawy, A.M.; El-Bassiouny, R.E.I.; Saleh, M.A. Browning inhibition mechanisms by cysteine, ascorbic acid and citric acid, and identifying PPO-catechol-cysteine reaction products. *J. Food Sci. Technol.* **2015**, *52*, 3651–3659. [CrossRef] [PubMed]
45. Oszmianski, J.; Lee, C.Y. Enzymic oxidative reaction of catechin and chlorogenic acid in a model system. *J. Agric. Food Chem.* **1990**, *38*, 1202–1204. [CrossRef]
46. Rico, D.; Martín-Diana, A.B.; Barat, J.M.; Barry-Ryan, C. Extending and measuring the quality of fresh-cut fruit and vegetables: A review. *Trends Food Sci. Technol.* **2007**, *18*, 373–386. [CrossRef]
47. Cadogan, E.I.; Lee, C.-H.; Popuri, S.R.; Lin, H.-Y. Effect of Solvent on Physico-Chemical Properties and Antibacterial Activity of Chitosan Membranes. *Int. J. Polym. Mater. Polym. Biomater.* **2014**, *63*, 708–715. [CrossRef]
48. Alvarez, M.V.; Ponce, A.G.; Moreira, M.D. Antimicrobial efficiency of chitosan coating enriched with bioactive compounds to improve the safety of fresh cut broccoli. *LWT Food Sci. Technol.* **2013**, *50*, 78–87. [CrossRef]
49. Karagoz, S.; Demirdoven, A. Effect of chitosan coatings with and without Stevia rebaudiana and modified atmosphere packaging on quality of cold stored fresh-cut apples. *LWT Food Sci. Technol.* **2019**, *108*, 332–337. [CrossRef]
50. Pushkala, R.; Parvathy, K.R.; Srividya, N. Chitosan powder coating, a novel simple technique for enhancement of shelf life quality of carrot shreds stored in macro perforated LDPE packs. *Innov. Food Sci. Emerg. Technol.* **2012**, *16*, 11–20. [CrossRef]

coatings

Article

Application of Pullulan and Chitosan Multilayer Coatings in Fresh Papayas

Linyun Zhang, Chongxing Huang * and Hui Zhao

School of Light Industry & Food Engineering, Guangxi University, Nanning 530004, China; 1716303004@st.gxu.edu.cn (L.Z.); zhh@gxu.edu.cn (H.Z.)
* Correspondence: huangcx@gxu.edu.cn; Tel.: +86-771-3272256

Received: 25 September 2019; Accepted: 8 November 2019; Published: 10 November 2019

Abstract: In this work, some multilayer coatings (two-layer, four-layer or six-layer) based on pullulan and chitosan for protecting papayas were prepared by the layer-by-layer technique. The papayas were coated by immersion and stored at 25 °C, 50% relative humidity or up to 14 days. Uncoated and monolayer-coated papayas were used as controls. The pullulan/chitosan coatings decreased the papaya weight loss, softening, color change (b^*, ΔE), and pH, retarded the fall of titratable acidity and vitamin C, and maintained respiratory rate and soluble solid contents. Sensory quality evaluation demonstrated that pullulan/chitosan coatings effectively preserved papaya flavor and overall acceptance. In general, the four-layer coatings provided the best fruit preservation. In conclusion, multilayer pullulan/chitosan coatings are efficient in maintaining the post-harvest quality and prolonging the shelf life of fresh papaya.

Keywords: papaya; preservation; pullulan; chitosan; multilayer coating

1. Introduction

Papaya is a rich resource of many active components with nutritional value such as vitamin C, carotene, and protease [1]. The papain found in papaya can be used to treat gastritis and indigestion, improve the nutritional value and functional properties of protein, and to produce shampoo, toothpaste, and beverages. Moreover, papayas have high medicinal and industrial value [2]. However, papayas are susceptible to fungal infections, such as anthracnose and stem-end rot, which are the major causes of papaya decay [3]. Papayas easily ripen at room temperature after harvest and quickly enter the senescence stage, with browning of the skin accompanied by dark spots and soft flesh [3]. Therefore, the development of an environmental and low-cost method for papaya preservation is necessary.

Several preservation technologies such as low temperatures treatment [4], hot water treatment [5], chemical treatments [6], and edible coatings [7] have been developed for papaya preservation. Among these approaches, edible coatings have become one of the major papaya preservation methods due to their relative low cost and simple application [8]. Polysaccharide-based coatings can control the internal atmosphere of fruits and delay ripening [9], as they provide a barrier against moisture, CO_2 and O_2 [10]. Zillo et al. [3] coated papayas with a carboxymethyl cellulose solution containing *Lippia sidoides* essential oils and measured the postharvest quality attributes of coated papayas. The results showed that the coating provided a good barrier to O_2 and water vapor, and extended the shelf life of papaya. However, essential oils used as antimicrobial agents are expensive and volatile.

Chitosan has broad-spectrum antimicrobial properties [11] as well as permselectivity to ethylene, CO_2 and O_2. This translates into a reduction in the respiration of fruit and might reduce the loss of organic substances [12]. However, chitosan film serves as a poor barrier for water vapor, which limits the application of chitosan in fruit preservation [13]. Priyadarshi et al. [14] prepared chitosan films containing different concentrations of apricot kernel essential oil (AKEO). The results showed that the incorporation of AKEO improved the resistance to moisture loss and water vapor barrier properties of the chitosan film. Pullulan is a nonionic polysaccharide with good adhesiveness and film formability [15]. Pullulan membranes are nearly impenetrable to gases such as O_2, N_2, CO_2 and aromas at low relative humidity (RH) [16]. Unfortunately, pullulan has barely any antifungal properties and is rarely used alone for fresh fruit preservation [17]. Silva et al. [16] incorporated lysozyme nanofibers into pullulan solutions to prepare a nanocomposite film. The nanofibers not only maintained the film-forming ability of pullulan but also imparted good antibacterial and mechanical properties to the nanocomposite film.

Multilayer coatings consist of different polymers that are alternately deposited on the target surface, and this preservation process is simple and inexpensive [18]. Multilayer coatings can combine the advantages of several single-layer coatings, with complete structure and stable performance [19]. Furthermore, multilayer coatings can effectively maintain the color, freshness, and firmness of fruit, thereby prolonging the postharvest life [20]. Yin et al. [21] deposited three-, five-, and seven-layer coatings based on chitosan solution containing cinnamon essential oil microcapsules and alginate solution on mango surface. The mangos were stored at 25 °C, 50% relative humidity (RH). The results indicated that multilayer coatings had dominant barrier effect to moisture and gases, maintained all the physiochemically indexes and extended the shelf life of mango effectively. Brasil et al. [22] applied a chitosan/pectin two-layer coating on the surface of fresh-cut papaya, and they added cinnamaldehyde microcapsules to the chitosan solution to enhance the antibacterial effect of the coating. The results showed that chitosan/pectin coating reduced the losses of vitamin C and total carotenoid content, and extended the shelf life of fresh-cut papaya stored at 4 °C for up to 15 d. However, available information on applications of pullulan/chitosan layer-by-layer coatings on entire fresh papaya is limited.

Hence, the aim of this study was to prepare multilayer coatings based on pullulan and chitosan and to evaluate their effectiveness for fresh papaya preservation.

2. Materials and Methods

2.1. Materials

Fresh papayas (*Carica papaya* L. cv 'Lingnanzhong') were purchased from a local fruit plantation in Nanning, Guangxi, China. The following reagents were used for this study: Chitosan (90.0% deacetylated, Sinopharm Chemical Reagent Co., Ltd., Shanghai, China); Glacial acetic acid (chemically pure, Tianjin Zhiyuan Chemical Reagent Co., Ltd., Tianjing, China); Pullulan (analytical pure, Aladdin Reagent Co., Ltd., Shanghai, China); Sodium hypochlorite solution (40% purity, Aladdin Reagent Co., Ltd., Shanghai, China); Sodium hydroxide (analytically pure, Aladdin Reagent (Shanghai) Co., Ltd., Shanghai, China); 2,6-dichlorophenolindophenol (analytical pure; Tianjin Guangfu Fine Chemical Research Institute); Oxalic acid (analytically pure, Tianjin Hengxing Chemical Preparation Co., Ltd., Tianjin, China); Kaolin (analytically pure, Tianjin Kemiou Chemical Reagent Co., Ltd., Tianjing, China). The pullulan and chitosan coatings are food grade, and can be used based on national and international regulations.

2.2. Preparation of Coating Solutions

The 0.5% (*w*/*v*) chitosan solution was prepared by dissolving 20 g of chitosan in 4 L 0.5% glacial acetic acid solution. The 0.5% (*w*/*v*) pullulan solution was prepared by dissolving 20 g of pullulan in 4 L of distilled water. The solutions were stirred for 3 h at room temperature and filtrated.

2.3. Coating of Fruits

Fresh papayas of the same size and color that had no bruises or black spots on the surface were selected, soaked in 0.3% (*v*/*v*) sodium hypochlorite solution for 3 min, rinsed with tap water for 2 min, and air-dried at room temperature.

The surface-disinfected papayas were randomly divided into six groups, with a control group and five test groups. Each group contained 30 papayas. Papayas were treated as described by Trevino-Garza et al. [20]. The treated papayas were dipped in 0.5% pullulan solution (*w*/*v*) for 5 min and the residual solution allowed to drip off for 2 min at 25 °C, which established the first layer on the surface of the fruit. The papayas were then immersed in 0.5% chitosan solution (*w*/*v*) for 5 min and allowed to stand at 25 °C for 2 min, which formed the second layer by hydrogen bonding between pullulan and chitosan. These steps were repeated to obtain the 4-layer and 6-layer multilayer coatings. The coated papayas were stored in a climate chamber (CLIMACELL404; Germany MMM Company, Planegg, Germany) at 25 °C and 50% RH. The physiological and nutritional attributes of papaya on days 0, 2, 4, 6, 8, 10, 12, and 14 were determined. For the experiment, 30 papayas were used per treatment during the entire storage period, and 3 fruits were used for quality mesurements per sampling day.

2.4. *Determination of Fruit Quality*

2.4.1. Weight loss, firmness, color and respiratory rate

A direct fruit weighing method described by Oregel-Zamudio et al. was used [23]. The papayas were weighed at each sampling time using an electronic scale (457A; Shengzhen Botoo Electronic Technology Co., Ltd., Shengzhen, China). The result was calculated as a percentage of total weight loss between the initial and final weights.

Firmness of papayas was measured as previously described by Jongsri et al. [24]. A fruit firmness meter (GY-2 type; Shanghai Hu Yueming Scientific Instrument Co., Ltd., Shanghai, China) was used to measure the firmness of each papaya by penetrating the skin with a Φ3.5 mm probe to a depth of 1 cm at three different locations on the fruit (proximal, distal, and middle).

As described by Peretto et al. [25], the color of papayas was determined using a spectrophoto-meter (CM-3600d; Japan Konica Minolta Company, Tokyo, Japan). The change in b^* value and ΔE was defined as color alteration. The formula used to determine ΔE is as follows:

$$\Delta E = \sqrt{\left(L_i^* - L_0^*\right)^2 + \left(a_i^* - a_0^*\right)^2 + \left(b_i^* - b_0^*\right)^2}, \quad (1)$$

where L_i^* represents the brightness value of papaya on day i, L_0^* represents the initial brightness value of papaya, a_i^* represents the red-green value of papaya on day i, a_0^* represents the initial red-green value of papaya, b_i^* represents the yellow-blue value of papaya on day i, and b_0^* represents the initial yellow-blue value of papaya.

Respiratory rate was measured using a flow-through system according to a method previously described by Gong et al. [26]. Papayas were individually weighed and placed in a fruit respiration apparatus (JFQ-315OH; Jun-Fang-Li-Hua Technology Institute, Beijing, China) and the valve was opened to allow air flow inside. Respiration was expressed as the CO_2 concentration ($mg \cdot kg^{-1} \cdot h^{-1}$).

2.4.2. Soluble Solids Content (SSC), Titratable Acidity (TA), pH and Vitamin C (VC) Content

The SSC of papayas was determined according to the method of Khaliq et al. [27]. First, 10 g of papaya pulp and 50 mL of distilled water were mixed to extract juice using a juicer (JYL-C020E; Nine Yang Co., Ltd., Shandong, China) for 3 min. After filtration, SSC in a drop of the supernatant was measured using an Abbe refractometer (WYA-2S; Shanghai Shen Guang Instrument Co., Ltd., Shanghai, China).

The TA of papayas was measured using a method previously described by Zhao et al. [28]. First, 10 g of papaya pulp were homogenized in 50 mL of distilled water using a juicer (JYL-C020E; Nine Yang Co., Ltd., Shandong, China) for 3 min and then filtered. The supernatant was incubated in water at 78 °C for 30 min. Next, 10 mL of mixture were titrated with 0.1 M NaOH. TA was calculated (%) with the formula below:

$$TA(\%) = \frac{\text{NaOH volume} \times 0.1 \times 0.064 \times 50\ \text{mL}}{10\ \text{g} \times 10\ \text{mL}}. \quad (2)$$

The pH of papayas was measured according to Temizkan et al. [29]. First, 10 g of papaya pulp was homogenized in 50 mL of distilled water using a juicer for 3 min and then filtered. The solution was collected and its pH was measured using a pH-meter (FE28; Jinan Guangyao Medical Equipment Co., Ltd., Shandong, China).

VC in papaya pulp was determined by 2,6-dichlorophenolindophenol (analytical pure; Tianjin Guangfu Fine Chemical Research Institute, Tianjin, China) titration [30]. First, 20 g of papaya pulp was mixed with 50 mL of 2% oxalic acid solution, and each sample was quickly homogenized using a juicer and filtered. Next, 20 mL of the filtrate were transferred to a 100 mL volumetric flask with 2% oxalic acid solution. Then, the solution was decolorized with 5 g of kaolin. The solution was filtered, and 10 mL of the filtrate was titrated with calibrated 2, 6-dichlorophenolindophenol until the solution turned pink and did not fade within 15 s. The VC content of papaya was calculated as follows:

$$VC = \frac{(V_i - V_0) \times T \times A}{m} \times 100, \quad (3)$$

where VC is the vitamin C content ($mg \cdot kg^{-1}$) of papaya, V_i is the volume (mL) of 2,6-dichlorophenolindophenol consumed in the titration of the sample, V_0 is the volume (mL) of 2,6-dichlorophenolindophenol consumed by blank titration, T is the titer ($mg \cdot mL^{-1}$) of 2,6-dichlorophenolate sodium, A is the dilution factor, and m is the papaya pulp weight (g).

2.4.3. Sensory Quality Evaluation

The appearance, flavor, and taste of the papayas were evaluated by 10 people who were professionally trained in sensory evaluation, as described by Ma et al. [31]. The score was based on a 10-point scale where a score of 5 was considered as the limit of acceptability. Scores below 5 corresponded to off-flavor production and a poor taste of papayas.

2.5. Statistical Analysis

The data were analyzed using an analysis of variance (ANOVA) with the SPSS 16.0 program (SPSS Inc., Chicago, IL, USA). Statistical significance was expressed at $P < 0.05$. The figures were drawn with Origin 8.1 (OriginLab Corp., Northampton, MA, USA).

3. Results and Discussion

3.1. Weight Loss

Papaya continuously loses moisture because of transpiration; therefore, the weight loss continues to increase. As shown in Figure 1a, irrespective of the treatment, weight loss of papayas increased as the storage time was prolonged. However, weight loss of uncoated papayas was significantly higher than that of coated papayas during storage ($P < 0.05$), as previously reported by others [3]. On the 14th day, weight loss of non-treated papayas reached 28.86%. The protective effects of coatings could be attributed to the semipermeable barrier to gas exchange and water loss created on the papaya's surface [32]. Compared with papayas coated with a single-layer, papayas coated with two, four or six layers showed lower weight loss (15.53%, 10.41% and 13.85%, respectively) on the 14th day. Pullulan and chitosan can form stable coatings on the papaya surface through hydrogen bonding and the coatings become denser as the layers increase; therefore, the barrier performance is better. Unfortunately, prolonged exposure of papaya to acidic chitosan solution may damage cell walls of the epidermis and lead to water loss of papaya, as shown in previous work with mango fruits [21]. In this work, the weight loss of the four-layer coated papayas was the smallest.

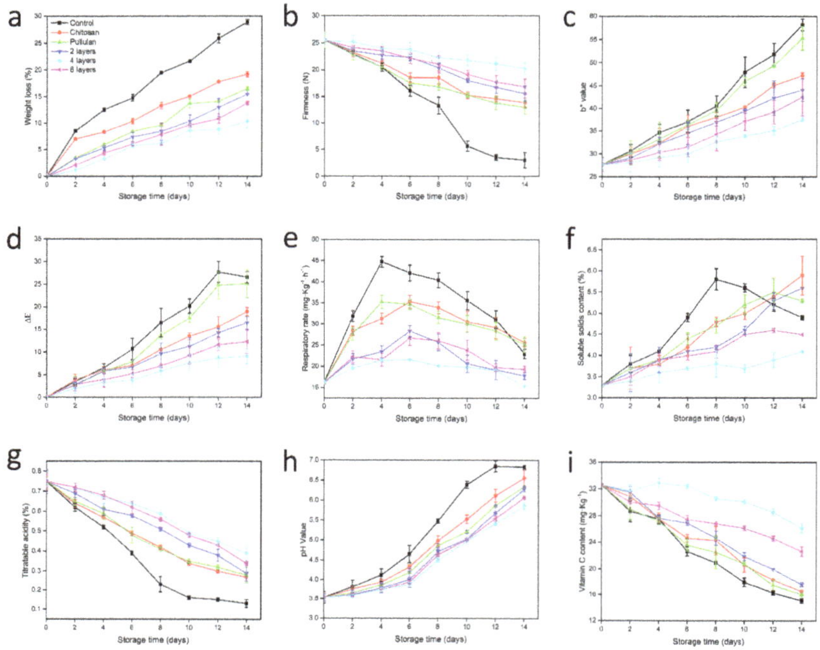

Figure 1. Effect of pullulan-chitosan multilayer coatings on (**a**) weight loss; (**b**) firmness; (**c**) b^* value; (**d**) ΔE; (**e**) respiratory rate; (**f**) SSC; (**g**) TA; (**h**) pH and (**i**) VC of papayas stored at 25 °C and 50% RH for 14 days. Vertical bars represent the standard errors of the means.

3.2. Firmness

During storage, pectin in papaya gradually decomposes by pectin degrading enzymes, which leads to a decrease in the firmness of papaya; the hydrolysis of starch in papaya can also promote fruit softening, as shown in other work with apples [33]. As shown in Figure 1b, the control samples exhibited a high rate of firmness loss compared with the coated samples ($P < 0.05$), as previously reported by others [34]. On day 14, uncoated control papayas were completely softened and could no longer be stored and transported. Papayas coated with two, four and six layers presented firmness values of 15.6, 20.2 and 16.9 N, respectively, on the 14th day, which were higher than the firmness of papayas coated with a single-layer, and this was consistent with prior research [32]. This result can be attributed to the low O_2 atmosphere created on the papaya's surface by the coating application, which can inhibit the activity of the enzymes involved in cell wall degradation processes and solubilization of pectins, as shown in previous work [35]. Moreover, the chitosan coating may also inhibited the activity of these enzymes [36]. As shown in Figure 1b, we also observed that the papayas coated with a six-layer coating were less firm than those coated with a four-layer coating, which may be due to the increased deposition time resulting in an adverse effect of the acidic chitosan solution to papaya epidermal cells [21]. In conclusion, the four-layer coating showed a positive effect in delaying fruit softening.

3.3. Color

Color is the most representative indicator of papaya maturity [37]. With prolonged storage time, chlorophyll is hydrolyzed by enzymatic action and decomposed by photooxidation, resulting in a gradual yellowing of papaya, as shown in other work with mangoes [38]. As shown in Figure 1c, the b^* value of the control samples increased rapidly, indicating that the papayas turned yellow quickly. On the 14th day, b^* value of uncoated papayas reached 58.1. However, the b^* value of coated papayas was lower and increased slower, as previously reported by others working with other fruits such as mango [39]. Papayas coated with two, four or six layers showed b^* values of 44.1, 37.6 and 42.6, respectively, all of which were lower than that of the single-layer coating ($P < 0.05$). The multilayer coating created a low O_2 atmosphere on the surface of papaya, inhibiting the activity of respiratory enzymes and ethylene synthetases [40]. These metabolic alterations delayed the fruit maturation process and maintained the stability of the cell wall for longer, which translated into greater protection to chloroplasts and consequently to chlorophylls [35]. In addition, the antimicrobial effects of the chitosan coating may also inhibit the activity of the chlorophyllase enzyme [41]. On the 14th day, the b^* value of four-layer coated papayas was 35.38% lower than that of uncoated papayas.

The ΔE value reflects the change in L^* and hue angle of the fruit [35]. The smaller the ΔE value, the more stable the color change of papaya [42]. As shown in Figure 1d, if compared with that of the control samples, the ΔE value of the coated fruits increased more slowly ($P < 0.05$), as shown in other work with mangoes [40]. Compared to papayas coated with other layers, the ΔE of papayas treated with a four-layer coating was minimal, and the color difference was more stable.

3.4. Respiratory Rate

Papaya is a climacteric fruit whose respiratory intensity changes with time and exhibits a sudden rise followed by a drop [43]. As shown in Figure 1e, uncoated papayas quickly reached the respiratory peak on the 4th day, and then the papaya respiratory rate rapidly decreased, as previously reported by Li et al. [44]. However, the respiratory peaks of multilayer coated papayas were both delayed until the 6th day, and the increase in respiration rate was suppressed compared with control fruits ($P < 0.05$). This indicates that the multilayer coating created a modified atmosphere with high CO_2 and low O_2 in the papaya, which reduced the respiratory rate, as shown in other work with mangoes [21]. Moreover, the antibacterial properties of the chitosan coating could have a role in the inhibition of the activity of respiratory enzymes and delayed respiration, as shown in previous work with nectarine fruits [45]. The respiratory rate of papaya with a four-layer coating was the most stable, better than that of papayas with a six-layer coating, which may be due to excessive barrier properties against O_2 of the six layer coating, resulting in anaerobic respiration of the papayas.

3.5. Soluble Solids Content

As shown in Figure 1f, the SSC of uncoated papayas increased quickly during the first 8 days of storage and then decreased rapidly, similar to that reported in previous studies with other fruits such as mango [46]. This may be because carbohydrates in fruits are hydrolysed into sugars during ripening, and this translates into an increase in SSC [47]. As the storage period was prolonged, the SSC reached the maximum value. Papaya respiration continued in the later period with strong microbial growth, resulting in a large amount of nutrient decomposition and a significant decrease in the SSC [48]. However, the change in SSC of coated papaya was observably lower than that in the control samples ($P < 0.05$). Compared with papayas coated with a single-layer, the SSC of papayas coated with two-layer, four-layer, and six-layer increased slower ($P < 0.05$). The multilayer coatings formed an effective barrier to O_2 in papaya, slowing down respiration and metabolic activity, and therefore retarded fruit ripening, as shown in other work with mangoes [21]. In fact, the change in SSC in the four-layer coated fruit group was minimal.

3.6. Titratable Acidity

TA content is regarded as an important indicator of respiration rate of fruits as organic acids are substrates for the respiratory metabolim [49]. As shown in Figure 1g, the TA content of coated and uncoated papayas decreased as the storage period was extended, as previously reported by others [49].

However, the TA of coated papayas showed a slower decrease than that of uncoated papayas ($P < 0.05$). Compared with single-layer coated papayas, at day 14, papayas coated with two-layer, four-layer, or six-layer coatings had higher TA (0.29%, 0.39% and 0.34%, respectively) ($P < 0.05$). Multilayer coatings reduced the respiratory rate of papayas, resulting in a decrease in the transformation of organic acids. Generally, four-layer coated papayas had higher TA at day 14.

3.7. pH

As shown in Figure 1h, pH values for all treatments increased continuously, which could be ascribed to a decline in the citric acid content of papayas [48]. However, by day 14 of storage, the pH values of coated papayas were lower than that of uncoated papayas ($P < 0.05$), which was consistent with previous results with other fruits such as mango [50]. This might be due to the gas barrier created by the coatings slowing down the metabolism of papaya and reducing the decomposition of organic acids [5]. On the 14th day, papayas coated with the four-layer coating had the lowest pH value.

3.8. Vitamin C

As shown in Figure 1i, the VC content of papayas declined during storage, although the coatings reduced the decrease ($P < 0.05$), as previously found by others [51]. After 14 days of storage, the VC content of control papayas decreased obviously, i.e., 15.06 mg·kg^{-1}, while papayas with a four-layer coating exhibited the highest VC content, i.e., 26.17 mg·kg^{-1} ($P < 0.05$). The multilayer coatings created a barrier to O_2 and CO_2, delaying the oxidation of VC [16]. The antimicrobial effects of the chitosan coating may have also inhibited the activity of ascorbate oxidase [45]. In conclusion, the application of ’he four-layer multilayer coating effectively delayed the oxidation of VC.

9. Sensory Quality Evaluation

As shown in Figure 2, the sensory quality of the control fruits decreased quickly. The control fruits their commercial value on the 8th day. This is mainly attributed to their high water loss and quick ning, causing bad appearance and poor taste of the papaya [30]. The sensory quality of coated as declined slowly, especially in case of the four-layer coated papayas, which still were rated points on the 14th day. This might be due to the effective barrier to O_2 and moisture of this It is worth mentioning that the sensory quality of the six-layer group decreased rapidly after 's until the fruits lost their commercial value, which might be due to the excessive ethanol by anaerobic respiration of the fruit, affecting the flavor and taste of papaya.

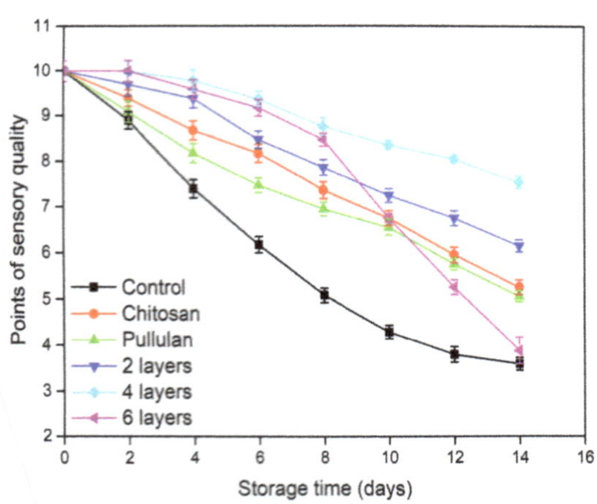

'y evaluation of papayas stored for 14 days at 25 °C, 50% RH. Scores below 5 ιot marketable. Vertical bars represent the standard errors of the means.

4. Conclusions

In summary, pullulan/chitosan multilayer coatings maintained the physiological and nutritional attributes of papayas stored at 25 °C and 50% RH, and extended the fruit shelf life if compared with uncoated and single-coated fruits. Sensory evaluation showed that the multilayer coatings maintained the flavor and commercial value of papayas for longer. This could be attributed to the ideal barrier to O_2 and moisture, and also to the inhibition of the activity of the respiratory enzymes. So pullulan/chitosan multilayer coatings could be applied as a new technique for fruit preservation. Among multilayer coatings, the best performance was obtained with the four-layer coating.

Author Contributions: Methodology, L.Z.; software, L.Z.; formal analysis, L.Z.; data curation, C.H.; writing—original draft preparation, L.Z.; writing—review and editing, H.Z.; visualization, H.Z.; supervision, C.H.

Funding: The authors are grateful for the financial support of the Guangxi Science and Technology Plan Project (Project No. 2018AB45007), the Opening Project of Guangxi Key Laboratory of Clean Pulp & Papermaking and Pollution Control (Project No. KF201607) and the Guangxi Natural Science Foundation Program (2018JJB120084).

Conflicts of Interest: The authors declare no conflict of interest. The sponsors had no role in the design, execution, interpretation, or writing of the study.

References

1. Yeoh, W.K.; Ali, A.; Forney, C.F. Effects of ozone on major antioxidants and microbial populations of fresh-cut papaya. *Postharvest Biol. Technol.* **2014**, *89*, 56–58. [CrossRef]
2. Lovera, N.N.; Ramallo, L.; Salvadori, V.O. Effects of different freezing methods on calcium enriched papaya (*Carica papaya* L.). *J. Food Sci. Technol.* **2018**, *55*, 2039–2047. [CrossRef] [PubMed]
3. Zillo, R.R.; Da Silva, P.P.M.; De Oliveira, J.; Da Gloria, E.M.; Spoto, M.H.F. Carboxymethylcellulose coating associated with essential oil can increase papaya shelf life. *Sci. Hortic.* **2018**, *239*, 70–77. [CrossRef]
4. Pan, Y.-G.; Yuan, M.-Q.; Zhang, W.-M.; Zhang, Z.-K. Effect of low temperatures on chilling injury in relation to energy status in papaya fruit during storage. *Postharvest Biol. Technol.* **2017**, *125*, 181–187. [CrossRef]
5. Promyou, S.; Supapvanich, S. Hot water incorporated with salicylic acid dips maintaining physicochemical quality of 'Holland' papaya fruit stored at room temperature. *Emir. J. Food Agric.* **2017**, *29*, 18–24.
6. Lata, D.; Aftab, M.A.; Homa, F.; Ahmad, M.S.; Siddiqui, M.W. Effect of eco-safe compounds on postharvest quality preservation of papaya (*Carica papaya* L.). *Acta Physiol. Plant.* **2018**, *40*, 8. [CrossRef]
7. Rangel-Marrón, M.; Mani-López, E.; Palou, E.; López-Malo, A. Effects of alginate-glycerol-citric acid concentrations on selected physical, mechanical, and barrier properties of papaya puree-based edible films and coatings, as evaluated by response surface methodology. *LWT* **2019**, *101*, 83–91. [CrossRef]
8. Salvador-Figueroa, M.; Castillo-Lopez, D.; Adriano-Anaya, L.; Galvez-Lopez, D.; Rosas-Quijano, R.; Vazquez-Ovando, A. Chitosan composite films: Physicochemical characterization and their use as coating in papaya Maradol stored at room temperature. *Emir. J. Food Agric.* **2017**, *29*, 779–791. [CrossRef]
9. Xu, F.; Liu, S.; Liu, Y.; Xu, J.; Liu, T.; Dong, S. Effectiveness of lysozyme coatings and 1-MCP treatments on storage and preservation of kiwifruit. *Food Chem.* **2019**, *288*, 201–207. [CrossRef]
10. Patel, C.; Panigrahi, J. Starch glucose coating-induced postharvest shelf-life extension of cucumber. *Food Chem.* **2019**, *288*, 208–214. [CrossRef]
11. Xing, K.; Li, T.J.; Liu, Y.F.; Zhang, J.; Zhang, Y.; Shen, X.Q.; Li, X.Y.; Miao, X.M.; Feng, Z.Z.; Peng, X.; et al. Antifungal and eliciting properties of chitosan against *Ceratocystis fimbriata* in sweet potato. *Food Chem.* **2018**, *268*, 188–195. [CrossRef] [PubMed]
12. Jiao, W.; Shu, C.; Li, X.; Cao, J.; Fan, X.; Jiang, W. Preparation of a chitosan-chlorogenic acid conjugate and its application as edible coating in postharvest preservation of peach fruit. *Postharvest Biol. Technol.* **2019**, *154*, 129–136. [CrossRef]
13. Halim, A.L.A.; Kamari, A.; Phillip, E. Chitosan, gelatin and methylcellulose films incorporated with tannic acid for food packaging. *Int. J. Biol. Macromol.* **2018**, *120*, 1119–1126. [CrossRef] [PubMed]
14. Priyadarshi, R.; Sauraj; Kumar, B.; Deeba, F.; Kulshreshtha, A.; Negi, Y.S. Chitosan films incorporated with Apricot (*Prunus armeniaca*) kernel essential oil as active food packaging material. *Food Hydrocoll.* **2018**, *85*, 158–166. [CrossRef]

15. Kraśniewska, K.; Ścibisz, I.; Gniewosz, M.; Mitek, M.; Pobiega, K.; Cendrowski, A. Effect of pullulan coating on postharvest quality and shelf-life of highbush blueberry (*Vaccinium corymbosum* L.). *Materials* **2017**, *10*, 965. [CrossRef]
16. Silva, N.H.; Vilela, C.; Almeida, A.; Marrucho, I.M.; Freire, C.S. Pullulan-based nanocomposite films for functional food packaging: Exploiting lysozyme nanofibers as antibacterial and antioxidant reinforcing additives. *Food Hydrocoll.* **2018**, *77*, 921–930. [CrossRef]
17. Wu, S.J.; Lu, M.S.; Wang, S.J. Effect of oligosaccharides derived from *Laminaria japonica*-incorporated pullulan coatings on preservation of cherry tomatoes. *Food Chem.* **2016**, *199*, 296–300. [CrossRef]
18. Arnon-Rips, H.; Poverenov, E. Improving food products' quality and storability by using Layer by Layer edible coatings. *Trends Food Sci. Technol.* **2018**, *75*, 81–92. [CrossRef]
19. Xiao, F.-X.; Pagliaro, M.; Xu, Y.-J.; Liu, B. Layer-by-layer assembly of versatile nanoarchitectures with diverse dimensionality: A new perspective for rational construction of multilayer assemblies. *Chem. Soc. Rev.* **2016**, *45*, 3088–3121. [CrossRef]
20. Arnon, H.; Granit, R.; Porat, R.; Poverenov, E. Development of polysaccharides-based edible coatings for citrus fruits: A layer-by-layer approach. *Food Chem.* **2015**, *166*, 465–472. [CrossRef]
21. Yin, C.; Huang, C.; Wang, J.; Liu, Y.; Lu, P.; Huang, L. Effect of chitosan- and alginate-based coatings enriched with cinnamon essential oil microcapsules to improve the postharvest quality of mangoes. *Materials* **2019**, *12*, 2039. [CrossRef] [PubMed]
22. Brasil, I.; Gomes, C.; Puerta-Gomez, A.; Castell-Perez, M.; Moreira, R.G. Polysaccharide-based multilayered antimicrobial edible coating enhances quality of fresh-cut papaya. *LWT* **2012**, *47*, 39–45. [CrossRef]
23. Oregel-Zamudio, E.; Angoa-Pérez, M.V.; Oyoque-Salcedo, G.; Aguilar-González, C.N.; Mena-Violante, H.G. Effect of candelilla wax edible coatings combined with biocontrol bacteria on strawberry quality during the shelf-life. *Sci. Hortic.* **2017**, *214*, 273–279. [CrossRef]
24. Jongsri, P.; Wangsomboondee, T.; Rojsitthisak, P.; Seraypheap, K. Effect of molecular weights of chitosan coating on postharvest quality and physicochemical characteristics of mango fruit. *LWT* **2016**, *73*, 28–36. [CrossRef]
25. Peretto, G.; Du, W.-X.; Avena-Bustillos, R.J.; Berrios, J.D.J.; Sambo, P.; McHugh, T.H. Electrostatic and conventional spraying of alginate-based edible coating with natural antimicrobials for preserving fresh strawberry quality. *Food Bioprocess Technol.* **2017**, *10*, 165–174. [CrossRef]
26. Gong, D.; Bi, Y.; Jiang, H.; Xue, S.; Wang, Z.; Li, Y.; Zong, Y.; Prusky, D. A comparison of postharvest physiology, quality and volatile compounds of 'Fuji' and 'Delicious' apples inoculated with *Penicillium expansum*. *Postharvest Biol. Technol.* **2019**, *150*, 95–104. [CrossRef]
27. Khaliq, G.; Mohamed, M.T.M.; Ding, P.; Ghazali, H.M.; Ali, A. Storage behaviour and quality responses of mango (*Mangifera indica* L.) fruit treated with chitosan and gum arabic coatings during cold storage conditions. *Int. Food Res. J.* **2016**, *23*, S141–S148.
28. Zhao, H.D.; Wang, B.G.; Cui, K.B.; Cao, J.K.; Jiang, W.B. Improving postharvest quality and antioxidant capacity of sweet cherry fruit by storage at near-freezing temperature. *Sci. Hortic.* **2019**, *246*, 68–78. [CrossRef]
29. Temizkan, R.; Atan, M.; Büyükcan, M.B.; Caner, C. Efficacy evaluation of ultrasound treatment on the postharvest storability of white nectarine by both physicochemical and image processing analyses. *Postharvest Biol. Technol.* **2019**, *154*, 41–51. [CrossRef]
30. Cissé, M.; Polidori, J.; Montet, D.; Loiseau, G.; Ducamp-Collin, M.N. Preservation of mango quality by using functional chitosan-lactoperoxidase systems coatings. *Postharvest Biol. Technol.* **2015**, *101*, 10–14. [CrossRef]
31. Ma, J.; Li, D.; Yang, D.; Xu, W.; Fu, Y.; Liao, R.; Shi, J.; Wang, J.; Wang, Y.; He, X. Effects of packaging designs with multiple pieces of function films on the quality of figs stored at ambient temperature. *Sci. Hortic.* **2019**, *251*, 32–38. [CrossRef]
32. Mendy, T.; Misran, A.; Mahmud, T.; Ismail, S. Application of *Aloe vera* coating delays ripening and extend the shelf life of papaya fruit. *Sci. Hortic.* **2019**, *246*, 769–776. [CrossRef]
33. Atkinson, R.G.; Sutherland, P.W.; Johnston, S.L.; Gunaseelan, K.; Hallett, I.C.; Mitra, D.; Brummell, D.A.; Schröder, R.; Johnston, J.W.; Schaffer, R.J. Down-regulation of POLYGALACTURONASE1 alters firmness, tensile strength and water loss in apple (*Malus* x *domestica*) fruit. *BMC Plant Biol.* **2012**, *12*, 129. [CrossRef] [PubMed]

34. Façanha, R.V.; Spricigo, P.C.; Purgatto, E.; Jacomino, A.P. Combined application of ethylene and 1-methylcyclopropene on ripening and volatile compound production of 'Golden' papaya. *Postharvest Biol. Technol.* **2019**, *151*, 160–169. [CrossRef]
35. Allanigue, D.K.A.; Sabularse, V.C.; Hernandez, H.P.; Serrano, E.P. The effect of chitosan-based nanocomposite coating on the postharvest life of papaya (*Carica papaya* L.) fruits. *Philipp. Agric. Sci.* **2017**, *100*, 233–242.
36. Wu, C.; Zhu, Y.; Wu, T.; Wang, L.; Yuan, Y.; Chen, J.; Hu, Y.; Pang, J. Enhanced functional properties of biopolymer film incorporated with curcurmin-loaded mesoporous silica nanoparticles for food packaging. *Food Chem.* **2019**, *288*, 139–145. [CrossRef]
37. Narsaiah, K.; Wilson, R.A.; Gokul, K.; Mandge, H.; Jha, S.; Bhadwal, S.; Anurag, R.K.; Malik, R.; Vij, S. Effect of bacteriocin-incorporated alginate coating on shelf-life of minimally processed papaya (*Carica papaya* L.). *Postharvest Biol. Technol.* **2015**, *100*, 212–218. [CrossRef]
38. Klangmuang, P.; Sothornvit, R. Active coating from hydroxypropyl methylcellulose-based nanocomposite incorporated with Thai essential oils on mango (cv. Namdokmai Sithong). *Food Biosci.* **2018**, *23*, 9–15. [CrossRef]
39. Zhang, B.; Huang, C.; Zhang, L.; Wang, J.; Huang, X.; Zhao, Y.; Liu, Y.; Li, C. Application of chlorine dioxide microcapsule sustained-release antibacterial films for preservation of mangos. *J. Food Sci. Technol.* **2019**, *56*, 1095–1103. [CrossRef]
40. Niu, B.; Shao, P.; Chen, H.; Sun, P. Structural and physiochemical characterization of novel hydrophobic packaging films based on pullulan derivatives for fruits preservation. *Carbohydr. Polym.* **2019**, *208*, 276–284. [CrossRef]
41. Romanazzi, G.; Feliziani, E.; Sivakumar, D. Chitosan, a biopolymer with triple action on postharvest decay of fruit and vegetables: Eliciting, antimicrobial and film-forming properties. *Front. Microbiol.* **2018**, *9*, 2745. [CrossRef] [PubMed]
42. Nowak, K.W.; Zielinska, M.; Waszkielis, K.M. The effect of ultrasound and freezing/thawing treatment on the physical properties of blueberries. *Food Sci. Biotechnol.* **2019**, *28*, 741–749. [CrossRef] [PubMed]
43. Hazarika, T.K.; Lalthanpuii; Mandal, D. Influence of edible coatings on physico-chemical characteristics and shelf-life of papaya (*Carica papaya*) fruits during ambient storage. *Indian J. Agric. Sci.* **2017**, *87*, 1077–1083.
44. Li, X.P.; Wu, B.; Guo, Q.; Wang, J.D.; Zhang, P.; Chen, W.X. Effects of nitric oxide on postharvest quality and soluble sugar content in papaya fruit during ripening. *J. Food Process. Preserv.* **2014**, *38*, 591–599. [CrossRef]
45. Zhang, W.; Zhao, H.; Zhang, J.; Sheng, Z.; Cao, J.; Jiang, W. Different molecular weights chitosan coatings delay the senescence of postharvest nectarine fruit in relation to changes of redox state and respiratory pathway metabolism. *Food Chem.* **2019**, *289*, 160–168. [CrossRef]
46. Chi, H.; Song, S.; Luo, M.; Zhang, C.; Li, W.; Li, L.; Qin, Y. Effect of PLA nanocomposite films containing bergamot essential oil, TiO_2 nanoparticles, and Ag nanoparticles on shelf life of mangoes. *Sci. Hortic.* **2019**, *249*, 192–198. [CrossRef]
47. Li, X.Y.; Du, X.L.; Liu, Y.; Tong, L.J.; Wang, Q.; Li, J.L. Rhubarb extract incorporated into an alginate-based edible coating for peach preservation Chock. *Sci. Hortic.* **2019**, *257*, 108685. [CrossRef]
48. Silva Neto, O.P.D.; Pinto, E.V.D.S.; Ootani, M.A.; Silva Junior, J.L.D.; Lima, J.L.D.S.B.; De Sousa, A.E.D. Ozone slows down anthracnose and increases shelf life of papaya fruits. *Rev. Bras. Frutic.* **2019**, *41*, 5. [CrossRef]
49. Escamilla-García, M.; Rodríguez-Hernández, M.J.; Hernández-Hernández, H.M.; Delgado-Sánchez, L.F.; García-Almendárez, B.E.; Amaro-Reyes, A.; Regalado-González, C. Effect of an edible coating based on chitosan and oxidized starch on shelf life of *Carica papaya* L., and its physicochemical and antimicrobial properties. *Coatings* **2018**, *8*, 318. [CrossRef]
50. Wang, J.; Wang, B.; Jiang, W.; Zhao, Y. Quality and shelf life of mango (*Mangifera indica* L. cv. 'Tainong') coated by using chitosan and polyphenols. *Food Sci. Technol. Int.* **2007**, *13*, 317–322. [CrossRef]
51. Silva, D.A.; Oliveira, J.K.; Santos, C.M.; Bery, C.C.S.; Castro, A.A.; Santos, J.A.B. The use of sodium alginate-based coating and cellulose acetate in papaya post-harvest preservation. *Acta Sci. Technol.* **2014**, *36*, 569–573. [CrossRef]

coatings

MDPI

Article

Effect of Nanostructured Chitosan/Propolis Coatings on the Quality and Antioxidant Capacity of Strawberries During Storage

Miriam del Carmen Martínez-González [1], Silvia Bautista-Baños [2], Zormy Nacary Correa-Pacheco [3], María Luisa Corona-Rangel [2], Rosa I. Ventura-Aguilar [3], Juan Carlos Del Río-García [4] and Margarita de Lorena Ramos-García [1,*]

1 Facultad de Nutrición, Calle Iztaccihuatl S/N, Universidad Autónoma del Estado de Morelos, Col. Los Volcanes, Cuernavaca, Morelos C.P. 62350, Mexico; miriam_martinez237@hotmail.com

2 Instituto Politécnico Nacional, Centro de Desarrollo de Productos Bióticos, Carretera Yautepec-Jojutla, km. 6.8, San Isidro, CEPROBI 8, Yautepec, Morelos C.P. 62731, Mexico; silviabb2008@hotmail.com (S.B.-B.); mlcr7@hotmail.com (M.L.C.-R.)

3 CONACYT-Centro de Desarrollo de Productos Bióticos, Instituto Politécnico Nacional, Carretera Yautepec-Jojutla, km. 6.8, San Isidro, CEPROBI 8, Yautepec, Morelos C.P. 62731, Mexico; zormynacary@yahoo.com (Z.N.C.-P.); iselaventura@yahoo.com.mx (R.I.V.-A.)

4 Facultad de Estudios Superiores, Universidad Nacional Autónoma de México, Campus 4, Cuautitlán Izcalli 54714, Mexico; mcjcrg@gmail.com

* Correspondence: margarita.ramosg@uaem.edu.mx

Received: 9 December 2019; Accepted: 17 January 2020; Published: 21 January 2020

Abstract: Strawberries have a thin epidermis and a high respiration rate. The use of edible coatings containing chitosan nanoparticles (CSNPs) and propolis (P) has been effective in preserving the shelf life and antioxidant capacity of various fruit and vegetable products. The present research evaluated the effect of coatings with CSNPs and P on the quality, antioxidant compounds, and antioxidant capacity of strawberries. The specific coatings that were evaluated were chitosan (CS), CS+CSNPs33%, CS + CSNPs + P10%, CS + CSNPs + P20%, CS + CSNPs + P30%, and a control with no coating. The variables were weight loss, firmness, total soluble solids (TSS), color, phenols, total flavonoids, antioxidant capacity, and sensory characteristics. An ANOVA and a Tukey test ($p \leq 0.05$) were used to analyze the data. Strawberries covered with CS + CSNPs + P10% showed the lowest weight loss (9.77%), while those covered with CS + CSNPs + P20% had the greatest firmness (4.96 N). CS + CSNPs + P coatings at 10%, 20%, and 30% concentrations maintained the antioxidant compounds and antioxidant capacity in the evaluated fruit (28.49 mg GAE g^{-1}, 554.61 μg quercetin g^{-1}, and 92.48% DPPH, respectively). The application of nanostructured coatings did not modify the sensory characteristics of the fruit. Coatings with CSNPs and/or P could therefore be a viable alternative for preserving the quality and antioxidant capacity of strawberries.

Keywords: nanoparticles; shelf life; *Fragaria* × *ananassa*; sensory characteristics; ripening

1. Introduction

The strawberry (*Fragaria* × *ananassa*) is one of the world's most popular fruit due to its excellent organoleptic and nutritional properties. The fruit has a very thin and fragile epidermis that makes it highly susceptible to mechanical damage during harvest and storage. This often results in a deterioration in quality, rapid loss of weight and firmness, and a loss of antioxidant capacity. In addition, the rate of respiration increases and there are changes in color [1]. The activity of enzymes such as polyphenol oxidase and peroxidase cause the degradation of anthocyanins and other polyphenols that lead to discoloration and increased darkening of the surface of the fruit [2]. The rate of deterioration of the fruit is proportional to the speed at which respiration occurs, approximately 15 mg kg^{-1} h^{-1} at 0 °C, increasing by 4–5 times when the temperature rises to 10 °C. This means that the fruit are highly perishable [3–5].

Edible coatings can retard the process of ripening, extend the shelf life, and prevent the loss of important components such as antioxidants in both fruit and vegetables [4]. For instance, chitosan (CS) has been used to improve the quality and extend the shelf life among others, of bananas, mango, guava, carambola, and figs [5–9]. However, it has been reported that a greater interaction on the surface area of the fruit can occur with the incorporation of chitosan nanoparticles into the coatings. In addition, the desired characteristics of the fruit, including, the quality attributes and the antioxidant capacity, also improved [10–12]. On this subject, Divya et al. (2018) [13] evaluated a coating containing chitosan nanoparticles at 5% in tomato and chilli pepper, and reported less weight (0.21% and 3.3% respectively). Also, Mohammadi et al. (2015) [14] evaluated chitosan nanoparticles at 0.15% in cucumber and found that the treated fruit had lower weight loss (9%) and CO_2 production (215 μg kg^{-1} s^{-1}), and higher firmness (55 N) compared to the control group (12%, 230 μg kg^{-1} s^{-1}, and 49 N, respectively). On this same line, Eshghi et al. (2014) [15] evaluated coatings with chitosan nanoparticles (0.25%) on various physicochemical characteristics and the functional profile on strawberries at 4 °C. The authors reported that, after eight days of storage, the coated strawberries showed lower weight (1.5%), firmness (27%), and respiration (33 mg kg^{-1} h^{-1}) than the remaining treatments. Chitosan can be combined with other hydrophobic compounds such as oils, waxes, and resins in order to improve the characteristics of the fresh, treated product [16]. For example, propolis extract is another natural product used to preserve the quality of fruit and vegetables. Barrera et al. (2012) [17] reported that papaya fruit covered with a commercial wax and propolis extract (5% w/v) had higher firmness (6.28 N) than fruit without propolis (5.4 N). In the same line, Ali et al. (2013) [18] found that a coating with propolis (5% *v/v*), gum arabic (5% *p/v*), and cinnamon oil (0.1% *v/v*) reduced the percentage of weight loss in chilli compared with the untreated vegetables (27% and 92%, respectively), and had the highest firmness (13 and 2 N, respectively). Also, Siripatrawan and Vitchayakitti (2016) [19] tested chitosan coatings combined with propolis extract (0, 2.5, 5, 10, and 20% *w/w*) and found that the total phenolic content and 1,1-diphenyl-2-picrilhydrazil (DPPH) in chitosan films increased due to the addition of propolis.

Although compounds such as chitosan and propolis have been shown to be effective in preserving the quality of various agricultural products, the combination of these compounds could show a synergic effect and improve the quality of the strawberries. Therefore, the objectives of this research were to evaluate the effect of coatings with CS, CSNP, and P on the ripening behavior and the antioxidant capacity of strawberries stored for a given time.

2. Materials and Methods

2.1. Materials

Strawberries cv. Camarosa were obtained from an orchard located in Tepoztlán, México (18°59′07″N 99°05′59″W). The fruit were harvested when 75% of the surface was red in accordance with NMX-FF-062-SCFI-2002 (Secretaría de Economía, 2002). Fruit with physical damage, irregular shapes, and the presence of microorganisms were discarded. A medium molecular weight chitosan Sigma Aldrich® (deacetylation degree 75%–85%) was used. The propolis extracts (10%, 20%, and 30%) were donated by the Laboratorio de Microbiología de la Unidad de Investigación Multidisciplinaria de la Facultad de Estudios Superiores (FES)-Cuautitlán, Universidad Autónoma de México. The ethanol was acquired from Hycel, Mexico, and the glacial acetic acid from Fermont Chemicals Inc, Mexico. The glycerol was purchased from J.T. Baker® (Randor, PA, USA)and Tween 20 from Meyer® (Tlahuac, Distrito Federal, Mexico).

2.2. Synthesis of Chitosan Nanoparticles

To obtain the CSNPs, the nanoprecipitation method reported by Correa-Pacheco [20] was followed. A CS solution (0.05% *w/v*) was dissolved in glacial acetic acid (1% *v/v*) and distilled water to form the aqueous phase. Then, 2.5 mL of this phase was added to 40 mL of the organic phase (ethanol) with 10 μL of Tween 20, using a peristaltic pump under constant magnetic stirring. The solution was placed in a rotary evaporator at 40 °C for solvent evaporation. The mean size of the CSNPs in ethanol was approximately 28.42 ± 7.43 nm, while the size of the CSNPs in propolis was 31.50 ± 7.77 nm, according to the results published previously by our research group [20]. Later, the obtained nanoparticles were stored under refrigeration at 4 °C and the concentration used in the coating was 33%.

2.3. Formulations and Application of Coatings

Five coatings were prepared: (1) CS, (2) CS + CSNPs, (3) CS + CSNPs + P10%), (4) CS + CSNPs + P20%, and (5) CS + CSNPs + P30%. The control consisted of dipping the fruit in water. The solution was homogenized with 1% chitosan (*w/v*), 1% acetic acid (*v/v*), and 0.3% glycerol (*v/v*), and the propolis extract was added by dripping using a peristaltic pump. The formulation was kept at 40 °C under constant stirring for 10 min and allowed to cool at room temperature. CSNPs were added to the formulation and stirring continued for another 5 min. The solution was then homogenized at 10,000 rpm for 1 min. The formulations were stored in amber colored bottles prior to use.

Strawberries were quickly washed with running water to remove excess dirt or garbage and allowed to dry, following which they were immersed for 30 s in each formulation, dried at room temperature, and stored in PET containers under refrigeration at 4 ± 1 °C for 8 days.

2.4. Determination of Weight Loss, Firmness, TSS, and Color

The fruit were weighed daily. Weight loss was determined by gravimetry with the help of a scale (OHAUS, Tokyo, Japan). This involved calculating the difference between the initial and final weight of each experimental unit, dividing this by the initial weight, and then multiplying the outcome by 100. The result was expressed as a percentage. Firmness was determined using an analogous penetrometer (KANDPI, Tokyo, Japan). A cylindrical tip 8 mm in diameter was used and both sides of the fruit were penetrated to a depth of 10 mm. Firmness was assessed at the beginning and end of the experiment. The values were reported as the force required to cross the membrane of the fruit in Newtons (N). To determine the TSS, a drop of strawberry juice was extracted and analyzed in a refractometer (ATAGO, Tokyo, Japan). The results were expressed in °Brix. The color of the fruit was determined daily using a colorimeter (Konica, Tokyo, Japan) for a period of 8 days. The CIEL * a * b * system values were evaluated. Measurements were made on both sides of the equatorial part of the fruit. Color values were reported in terms of the coordinate's luminosity (L *), hue angle (H * = tan − 1 b * / a *), and chromaticity (C * = √ (a *) 2 + (b *) 2). The CIELAB data were transformed to RGB values, using the nix color sensor converter.

2.5. Total Phenolic Compounds

The quantification of total phenolic compounds was evaluated every third day and analyzed using the Folin–Ciocalteau colorimetric method described by Singleton and Rossi (1965) [21]. Subsequently, 150 μL of the sample was mixed with 3.85 mL of distilled water, 250 μL of Folin–Ciocalteau phenol, and 750 μL of $NaCO_3$. These were allowed to react for 2 h in the dark at room temperature. Absorbance was measured at 760 nm with a spectrophotometer (Thermo scientific Genesys, Shanghai, China) and the concentration of total phenolic compounds in the samples was expressed as mg of gallic acid equivalents.

2.6. Total Flavonoids

To quantify the flavonoid content, the methodology by Chougui et al. (2013) [22] was followed. Briefly, Two g of strawberry were macerated with 5 mL of an 80% methanol solution and centrifuged at 8000 rpm for 12 min. Then, 1.5 mL of supernatant was then added, which reacted with 1.5 mL of $AlCl_3$. After 30 min, the absorbance of the sample was read at 430 nm with a spectrophotometer (Thermo scientific Genesys, China). Flavonoid content was quantified using a standard quercetin curve (20–110 μg quercetin) and evaluated every third day.

2.7. Antioxidant Capacity

The isolation and determination of extracts rich in bioactive compounds, enables the identification of the antioxidant properties, and this information can be considered as an indicator of the antioxidant properties of a food matrix. (Durazo and Lucarini, 2019) [23]. To quantify the antioxidant capacity, the methodology employed by Brand-Williams (1995) [24] was followed. Here, 0.01g of DPPH was weighed and added to 25 mL with methanol. Then, 10 mL with methanol was then added to 1.3 mL of the solution to prepare the daily solution. Subsequently, 0.5 g of the strawberry sample were weighed, following which 5 mL of methanol was added, macerated with a ceramic mortar, and centrifuged (Labnet International, New York, NJ, USA) at 800 rpm for 10 min. 250 μL of the sample was then taken and added to 750 μL of DPPH (133μM). For the blank, 750 μL of DPPH was added to 250 μL of methanol. The sample was incubated in the dark at room temperature for 30 min. Absorbance at 517 nm was then measured (Thermo scientific Genesys, Shanghai, China). Radical uptake activity was expressed as a percentage of DPPH inhibition and was calculated according to the following formula:

$$\%\ reduccion\ of\ DPPH = Abs0 - Absm\ \times 100\ \div Abs0$$

where Abs0* denotes blank absorbance and Absm**, sample absorbance.

2.8. Sensory Evaluation

The sensory evaluation was carried out on strawberries covered with the five different treatments. Five strawberries were used per treatment, including the control. Covered strawberries were cut in half and those coded with random digits were then placed in white plastic cups. The glasses were closed for 20 min. 30 untrained judges evaluated two random samples in which aroma, color, and flavor were rated on a scale from 1 to 9, whereby 1 means "I extremely dislike it" and 9 "I extremely like it". The judges ate a salty cookie between each sample so that the first sample did not influence the second.

2.9. Statistical Analysis

An analysis of variance (ANOVA) and a Tukey means test ($p \leq 0.05$) were then performed, using the statistical package InfoStat student version 2018. Fifteen treated strawberries with 3 repetitions were used in the variables of weight, firmness, TSS, color, total phenols, total flavonoids and antioxidant capacity, while 10 samples per treatment were used in the sensory evaluation.

3. Results and Discussion

For all treated and non-treated fruit, there was a continuous loss of weight during the eight days of storage (Table 1). The strawberries with the lowest weight at the end of the storage period were those coated with CS + CSNPs + P10% (9.7 and 10.2%, respectively) followed by the coatings CS + CSNPs + P30% and CS + CSNPs + P20% (10.9 and 11.0%, respectively). There was a significant difference ($p \leq 0.05$) between these and the controls (14.9%). These results aligned with those reported by Gol et al. (2013) [25]. A lower percentage of weight loss (4.0%) in strawberries coated with Hydroxypropylmethylcellulose 1% (HPMC) and CS 1% was obtained with respect to the control (14.3%) after eight days of storage. This may have occurred because the edible coatings served as a semipermeable barrier between the oxygen, carbon dioxide, and moisture loss, which reduced respiration, water loss, and oxidation reactions [26].

Table 1. Weight loss of strawberries treated with nanostructured chitosan coatings and propolis extract during eight days of storage.

	Weight Loss (%)					
	Storage Days					
Coatings	1	2	3	4	5	8
CS	0.0	1.5 ± 1.5 ^{ab*z**}	4.3 ± 2.0 ^{bcy}	7.6 ± 2.0 ^{bcx}	9.2 ± 2.3 ^{ax}	13.2 ± 2.9 ^{bcw}
CS + CSNPs	0.0	0.5 ± 1.5 ^{az}	2.9 ± 2.1 ^{aby}	5.6 ± 2.3 ^{abx}	7.4 ± 2.5 ^{abx}	10.2 ± 2.7 ^{aw}
CS + CSNPs + P10%	0.0	1.2 ± 1.7 ^{abzy}	2.5 ± 1.5 ^{ay}	4.5 ± 1.9 ^{ax}	6.6 ± 1.8 ^{aw}	9.7 ± 1.8 ^{av}
CS + CSNPs + P20%	0.0	0.9 ± 0.4 ^{abz}	3.3 ± 1.1 ^{aby}	5.9 ± 1.3 ^{abx}	7.8 ± 1.4 ^{abw}	11.0 ± 1.8 ^{abv}
CS + CSNPs + P30%	0.0	1.1 ± 0.6 ^{abz}	3.6 ± 1.3 ^{abcy}	6.6 ± 1.9 ^{abx}	8.0 ± 1.7 ^{abx}	10.8 ± 2.1 ^{abw}
Control	0.0	1.94 ± 0.7 ^{bz}	5.1 ± 1.5 ^{cy}	9.1 ± 3.0 ^{cx}	12.1 ± 3.1 ^{cw}	14.9 ± 2.8 ^{cv}

* Means with similar letters (a, b and c) are not significantly different among the evaluated treatments. ** Means with similar letters (w, x, y and z) are not significantly different among the storage days. CS (chitosan), CSNPs (chitosan nanoparticles), P10%, 20%, and 30% (propolis extract at different concentrations), control (strawberry fruit without the edible coating). Fifteen treated strawberries with 3 repetitions were used and an ANOVA and Turkey test ($p \leq 0.05$) were performed.

After eight days storage, the strawberries coated with CS + CSNPs + P20% and CS + CSNPs + P10% showed the greatest firmness (4.96 N and 4.87 N, respectively). These two values were statistically similar to the rest of the treatments but significantly different ($p \leq 0.05$) from the control (3.83 N) (Table 2). These results align with those of Restrepo et al. (2010) [27] who reported greater firmness in strawberries covered with mucilaginous gel of aloe penca and carnauba wax. They also reported the lowest firmness and highest weight loss in non-coated strawberries. Ventura-Aguilar et al. (2018) [28] evaluated the effect of a chitosan and cinnamon essential oil coating applied to strawberries at 5 °C and 20 °C. The results indicated that weight loss was reduced by 15 times and firmness was 33% higher in the fruit treated with the coating compared with the control. By contrast, studies carried out by Pilon et al. (2014) [29] reported no significant differences in firmness values between freshly cut apples covered with chitosan nanoparticles and uncovered apples. The coatings with CSNPs and the control group showed statistical differences among the storage days.

Table 2. Firmness values of strawberry fruit treated with nanostructured chitosan coatings and propolis extract during eight days of storage.

Coatings	Firmness (N)	
	Storage Days	
	1	8
CS	5.45 ± 0.97 ^{a*z**}	4.82 ± 1.04 ^{abz}
CS + CSNPs	5.43 ± 1.00 ^{ay}	4.50 ± 0.94 ^{abz}
CS + CSNPs + P10%	5.15 ± 1.05 ^{az}	4.87 ± 0.67 ^{bz}
CS + CSNPs + P20%	5.19 ± 0.93 ^{az}	4.96 ± 1.26 ^{bz}
CS + CSNPs + P30%	5.24 ± 1.28 ^{az}	4.75 ± 1.02 ^{abz}
Control	5.31 ± 0.91 ^{ay}	3.83 ± 0.75 ^{az}

* Means with similar letters (a and b) are not significantly different among the evaluated treatments. ** Means with similar letters (y and z) are not significantly different among the storage days. CS (chitosan), CSNPs (chitosan nanoparticles), P10%, 20%,and 30% (propolis extract at different concentrations). Fifteen treated strawberries with 3 repetitions were used and an Anova and tukey test ($p \leq 0.05$) were performed.

In this research, the values obtained for the weight and firmness of the strawberries could be due to the combination of propolis and chitosan extract. For instance, Bodini et al. (2013) [30] found that incorporating propolis extract (5%) significantly reduced the permeability of water vapor in relation to a control film (2.4 and 3.2 g mm/h cm^2 Pa, respectively). Similarly, Siripatrawan and Vitchayakitti

(2016) [19] studied the effect of propolis (2%) on the functional properties of chitosan films and found that this reduced the permeability of water vapor in the films (0.5 g mm Pa^{-1} d^{-1} m^2). This was because the polyphenolic compounds of the propolis stuck to the chitosan matrix and engaged in hydrogen or covalent bonding with chitosan reactive groups. This limited the availability of hydrogen atoms needed to form a hydrophilic bond with water. This eventually led to a decrease in the affinity of chitosan films towards water, and thus reduced the water vapor permeability of the coatings.

With respect to the TSS, there were no significant statistical differences between the treatments (Table 3). This aligns with the findings of Pastor et al. (2010) [31], who tested an edible coating made of HPMC (5%) and propolis extract (0.5%, 1%, and 1.5%) on grapes cv. Muscatel and found no significant differences. Similarly, Barrera et al. (2012) [17] applied propolis extract (5% *w/v*) to papaya fruit and found no significant differences with respect to the TTS. In general, statistical differences were not observed between the storage days of the evaluated fruit.

Table 3. TSS content of strawberry fruit treated with nanostructured chitosan and propolis extract coatings during eight days of storage.

Coatings	TSS (°BRIX)		
	Storage Days		
	1	5	8
CS	6.72 ± 0.81 ^{a*z**}	6.85 ± 0.84 ^{az}	6.79 ± 0.65 ^{az}
CS + CSNPs	6.68 ± 0.64 ^{az}	6.83 ± 0.72 ^{az}	6.87 ± 0.74 ^{az}
CS + CSNPs + P10%	6.67 ± 0.77 ^{az}	6.77 ± 0.68 ^{az}	6.83 ± 1.02 ^{az}
CS + CSNPs + P20%	6.61 ± 0.70 ^{az}	6.67 ± 0.82 ^{az}	6.71 ± 1.13 ^{az}
CS + CSNPs + P30%	6.93 ± 0.88 ^{az}	6.80 ± 0.68 ^{az}	6.78 ± 1.01 ^{az}
Control	6.82 ± 0.82 ^{az}	7.05 ± 1.21 ^{az}	6.71 ± 0.76 ^{az}

* Means with similar letters (a) are not significantly different among the evaluated treatments. ** Means with similar letters (z) are not significantly different among the storage days. CS (chitosan), CSNPs (chitosan nanoparticles), P10, 20, and 30% (propolis extract at different concentrations). Fifteen treated strawberries with 3 repetitions were used and an ANOVA and Turkey test ($p \leq 0.05$) were performed.

Regarding luminosity, chromaticity, and hue angle, no significant statistical differences ($p \leq 0.05$) were observed between the treatments (Figure 1).

Trejo et al. (2007) [32] reported that the application of an edible coating using 1% and 2% gelatin did not affect the luminosity of strawberries stored in refrigeration for 10 days. Similarly, Del Valle et al. (2005) [33] reported no changes in color parameters for strawberries coated with cactus mucilage and glycerol stored for 10 days at 5 °C. The color of the strawberry is a very important attribute in the acceptance of the product by the consumer, and the edible coatings applied in this study did not modify the original color.

The total phenol content decreased during the storage period for all treatments. However, strawberries coated with formulations containing chitosan and propolis exhibited the highest phenolic content on each day of storage (Table 4). On the first day, the corresponding values were 69.98, 67.15, and 66.46 mg GAE g^{-1} strawberry, respectively. On the third day of storage, the fruit coated with CS + CSNPs + P20% showed the highest content of total phenolic compounds (45.49 mg GAE g^{-1} strawberry) compared with the control and the remaining treatments. During the following two sampling periods, the highest content was in the fruit coated with CS + CSNPs + P10% (34.75 mg GAE g^{-1} and 28.49 mg GAE g^{-1} strawberry, respectively), which was significantly different ($p \leq 0.05$) from the control (18.13 mg GAE g^{-1} strawberry). However, it was statistically similar to the other coatings tested. In all treatments, significant differences were observed and a decrease of phenol content was observed at the end of the storage time.

Luminosity (L*)

a)

Coatings	Storage days 1	2	3	4	5	8
CS	34.09 ± 2.46	33.14 ± 3.19^{a}	33.90 ± 1.88^{ab}	30.29 ± 2.20^{a}	31.56 ± 1.87^{a}	30.98 ± 2.28^{a}
CS + CSNPs	32.95 ± 2.04^{a}	33.28 ± 2.84^{a}	32.89 ± 3.12^{ab}	31.55 ± 2.30^{a}	32.51 ± 2.49^{a}	31.54 ± 2.41^{a}
CS + CSNPs + P10%	34.84 ± 2.42^{a}	32.19 ± 2.19^{a}	33.15 ± 2.33^{ab}	32.24 ± 1.48^{a}	31.32 ± 2.08^{a}	31.77 ± 1.53^{a}
CS + CSNPs + P20%	33.48 ± 3.74^{a}	31.65 ± 3.63^{a}	32.21 ± 2.88^{a}	30.90 ± 2.49^{a}	30.18 ± 2.83^{a}	30.64 ± 2.54^{a}
CS + CSNPs + P30%	32.49 ± 4.43^{a}	31.50 ± 3.26^{a}	32.98 ± 2.29^{ab}	31.92 ± 2.99^{a}	30.95 ± 1.91^{a}	31.17 ± 2.27^{a}
Control	33.73 ± 1.83^{a}	34.42 ± 3.41^{a}	35.45 ± 3.12^{b}	32.00 ± 2.86^{a}	32.56 ± 2.60^{a}	32.32 ± 2.62^{a}

Chromaticity (C*)

b)

Coatings	Storage days 1	2	3	4	5	8
CS	27.9 ± 3.3^{a}	30.3 ± 4.3^{a}	26.0 ± 2.4^{abc}	29.7 ± 3.2^{b}	27.3 ± 2.7^{ab}	26.8 ± 2.8^{b}
CS + CSNPs	30.3 ± 3.3^{abc}	31.3 ± 3.7^{a}	27.4 ± 3.2^{bc}	28.1 ± 2.7^{ab}	27.0 ± 2.9^{ab}	27.3 ± 2.2^{b}
CS + CSNPs + P10%	32.6 ± 4.9^{bc}	30.5 ± 3.5^{a}	26.0 ± 2.2^{abc}	28.9 ± 4.3^{ab}	27.8 ± 2.5^{ab}	27.3 ± 3.2^{b}
CS + CSNPs + P20%	33.6 ± 3.1^{c}	28.6 ± 4.4^{a}	23.7 ± 3.1^{a}	27.1 ± 4.4^{ab}	25.8 ± 3.9^{a}	22.8 ± 3.8^{a}
CS + CSNPs + P30%	31.6 ± 3.2^{abc}	29.9 ± 3.9^{a}	24.9 ± 3.0^{ab}	26.0 ± 2.0^{a}	26.5 ± 2.5^{ab}	25.6 ± 2.6^{ab}
Control	29.7 ± 2.9^{ab}	30.3 ± 3.2^{a}	28.8 ± 2.9^{c}	30.0 ± 2.3^{ab}	29.0 ± 2.0^{b}	28.0 ± 2.8^{b}

Hue angle (H°)

c)

Coatings	Storage days 1	2	3	4	5	8
CS	25.5 ± 4.1^{a}	25.0 ± 5.0^{a}	20.6 ± 4.1^{a}	24.7 ± 5.4^{a}	21.5 ± 4.0^{a}	22.8 ± 3.9^{a}
CS + CSNPs	25.1 ± 6.4^{a}	24.0 ± 4.2^{a}	21.7 ± 3.8^{a}	23.8 ± 3.9^{a}	21.0 ± 3.8^{a}	22.7 ± 3.8^{a}
CS + CSNPs + P10%	27.2 ± 9.1^{a}	23.1 ± 3.7^{a}	20.8 ± 3.7^{a}	23.1 ± 3.5^{a}	21.5 ± 2.4^{a}	22.6 ± 2.9^{a}
CS + CSNPs + P20%	28.7 ± 5.7^{a}	24.3 ± 3.6^{a}	20.7 ± 2.9^{a}	24.4 ± 7.4^{a}	21.8 ± 3.5^{a}	22.5 ± 3.1^{a}
CS + CSNPs + P30%	26.7 ± 4.8^{a}	24.6 ± 6.6^{a}	21.2 ± 4.2^{a}	22.7 ± 5.3^{a}	22.4 ± 3.8^{a}	21.9 ± 3.9^{a}
Control	25.0 ± 5.9^{a}	25.3 ± 5.4^{a}	23.7 ± 4.2^{a}	23.9 ± 4.4^{a}	23.0 ± 4.1^{a}	23.8 ± 5.0^{a}

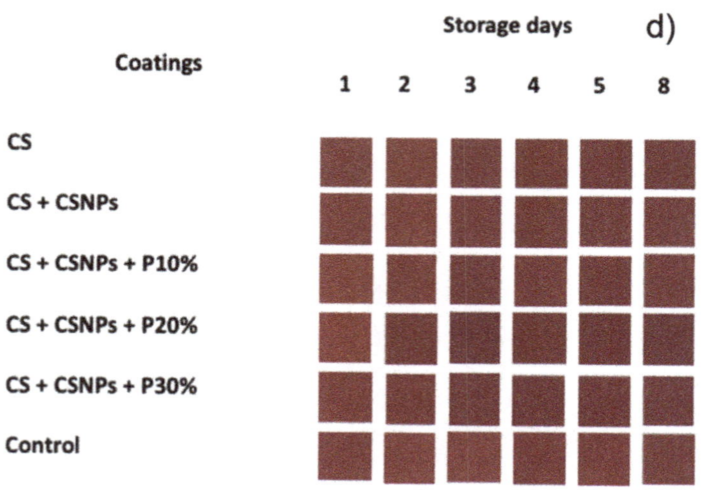

Figure 1. Change in color of strawberry fruit treated with nanostructured chitosan and propolis extract coatings during eight days of storage; (**a**) Luminosity values of strawberry fruit treated with nanostructured chitosan and propolis extract coatings. (**b**) Chromaticity values of strawberry fruit treated with nanostructured chitosan and propolis extract coatings. (**c**) Hue angle values of strawberry

fruit treated with nanostructured chitosan and propolis extract coatings. (**d**) Squares of color determined through the coordinates L *, a *, b *, and transformed to RGB values by Nix color sensor. Fifteen treated strawberries with 3 repetitions were used and an Anova and tukey test ($p \leq 0.05$) were performed. CS (chitosan), CSNPs (chitosan nanoparticles), P10%, 20%, and 30% (propolis extract at different concentrations). Control (strawberry fruit without the edible coating). Means with equal letters are not significantly different. ANOVA and Turkey test ($p \leq 0.05$).

Table 4. Total phenol content of strawberry fruit treated with nanostructured chitosan and propolis coatings during eight days of storage.

Coatings	Total Phenols (μg GAE g^{-1} of Strawberry)			
	Storage Days			
	1	3	5	8
CS	55.07 ± 4.58 ^{ab*x**}	41.69 ± 1.54 ^{bcy}	31.18 ± 4.59 ^{abz}	26.23 ± 3.52 ^{abz}
CS + CSNPs	57.80 ± 4.73 ^{bcx}	38.49 ± 2.10 ^{by}	33.27 ± 5.74 ^{aby}	24.19 ± 5.80 ^{abz}
CS + CSNPs + P10%	66.46 ± 2.46 ^{cdx}	39.10 ± 1.95 ^{by}	34.75 ± 2.32 ^{by}	28.49 ± 1.20 ^{bz}
CS + CSNPs + P20%	67.15 ± 2.65 ^{dx}	45.49 ± 0.56 ^{cy}	31.83 ± 3.22 ^{abz}	25.53 ± 2.63 ^{abz}
CS + CSNPs + P30%	69.98 ± 2.02 ^{dx}	40.12 ± 0.77 ^{by}	30.67 ± 4.54 ^{abz}	24.47 ± 1.35 ^{abz}
Control	48.08 ± 0.66 ^{ay}	24.4 ± 2.05 ^{az}	22.85 ± 4.01 ^{az}	18.13 ± 2.92 ^{az}

* Means with similar letters (a, b, c and d) are not significantly different among the evaluated treatments. ** Means with similar letters (x, y and z) are not significantly different among the storage days. CS (chitosan), CSNPs (chitosan nanoparticles), P10, 20, and 30% (propolis extract at different concentrations). Fifteen treated strawberries with 3 repetitions were used and an Anova and tukey test ($p \leq 0.05$) were performed.

The total flavonoid content was generally reduced during storage (Table 5). For each storage evaluation, fruit from the treatments CS + CSNPs and CS + CSNPs + P20% showed the highest flavonoid content. The corresponding values were in the range of 954 to 554.6 μg quercetin g^{-1} and were significantly different ($p \leq 0.05$) from those of the other treatments. In all treatments, significant differences were observed and a decrease of flavonoids content was observed at the end of the storage.

Table 5. Total flavonoid content of strawberry fruit treated with nanostructured chitosan and propolis coatings during eight days of storage.

Coatings	Total Flavonoids (μg Quercetin g^{-1})			
	Storage Days			
	1	3	5	8
CS	667.1 ± 5.8 ^{b*x**}	750.5 ± 11.6 ^{dw}	515.8 ± 9.7 ^{by}	409.1 ± 20.2 ^{az}
CS + CSNPs	698.1 ± 8.4 ^{cw}	620.5 ± 6.7 ^{bx}	522.9 ± 11.6 ^{by}	402.6 ± 4.0 ^{az}
CS + CSNPs + P10%	853.7 ± 4.4 ^{ew}	667.1 ± 10.0 ^{cx}	546.8 ± 17.4 ^{by}	505.4 ± 11.6 ^{bz}
CS + CSNPs + P20%	954.2 ± 8.8 ^{fw}	974.9 ± 15.8 ^{ex}	645.1 ± 9.9 ^{cy}	554.6 ± 3.3 ^{cz}
CS + CSNPs + P30%	758.9 ± 5.6 w	647.0 ± 15.0 ^{bcx}	400.7 ± 16.2 ^{az}	498.3 ± 3.8 ^{by}
Control	588.2 ± 4.4 ^{ax}	459.5 ± 1.9 ^{az}	524.2 ± 11.2 ^{by}	508.0 ± 13.9 ^{by}

* Means with similar letters (a, b, c, d, e and f) are not significantly different among treatments. ** Means with similar letters (w, x, y and z) are not significantly different among the storage days. CS (chitosan), CSNPs (chitosan nanoparticles), P10, 20, and 30% (propolis extract at different concentrations). Fifteen treated strawberries with 3 repetitions were used and an Anova and tukey test ($p \leq 0.05$) were performed.

The percentage of DPPH inhibition was higher in the coated strawberries compared with the control regardless of the applied treatment (Table 6). In general, strawberries coated with CS, CSNPs, and P at different concentrations had the highest percentage of DPPH inhibition with values of 87.5, 90.2%, 79.8%, and 92.4%, respectively. These results are consistent with those reported by Wang and Gao (2013) [34] who demonstrated that the ability to sequester free radicals in chitosan-coated strawberries cv Earliglow was higher than that of uncoated fruit at the end of nine days of storage at 5 °C. Similarly, López et al. (2012) [35] evaluated the effect of a coating of CS (1%) and cinnamon essential oil (0.03%) on preserving the quality, antioxidant capacity, and phenolic content of strawberries stored at 5 °C. They found that the coating was effective for maintaining the total phenolic content (170 mg GAE g^{-1}) and antioxidant capacity (inhibition 85% DPPH) for up to 15 days. Zahid et al. (2013) [36] also reported an increase in the biosynthesis of antioxidants and total flavonoids following the application of 0.5% propolis extract. According to Thomas et al. (2016) [37] and Anjum et al. (2013) [38], the propolis has high biological activity, including antioxidant activity, due to its high phenolic and flavonoid content. In all treatments, significant statistical differences were observed and a decrease in DPPH inhibition was observed at the end of the storage.

Table 6. Antioxidant capacity of strawberry fruit treated with nanostructured chitosan and propolis coatings during eight days of storage.

Coatings	DPPH (%)			
	Storage Days			
	1	3	5	8
CS	86.0 ± 3.6 ^{ab*y**}	73.9 ± 6.5 ^{az}	79.8 ± 1.6 ^{czy}	83. 6 ± 2.3 ^{bcxy}
CS + CSNPs	87.5 ± 1.9 ^{by}	78.1 ± 7.0 ^{abzy}	72.6 ± 0.8 ^{bz}	78.2 ± 3.0 ^{bzy}
CS + CSNPs + P10%	79.0 ± 2.1 ^{aby}	82.5 ± 1.1 ^{abyz}	73.9 ± 2.4 ^{bcz}	87.0 ± 1.5 ^{bcx}
CS + CSNPs + P20%	87.3 ± 5.8 ^{by}	80.8 ± 8.3 ^{abzy}	72.6 ± 4. 3 ^{bz}	83.5 ± 0.5 ^{byz}
CS + CSNPs + P30%	82.3 ± 5.8 ^{abzy}	90.2 ± 0.3 ^{by}	73.7 ± 0.9 ^{bz}	92.4 ± 6.6 ^{cy}
Control	75.6 ± 0.2 ^{ax}	69.5 ± 1.9 ^{ay}	57.3 ± 0.4 ^{ay}	69.0 ± 0.6 ^{az}

* Means with similar letters (a, b and c) are not significantly different among the evaluated treatments. ** Means with similar letters (x, y and z) are not significantly different among the storage days. CS (chitosan), CSNPs (chitosan nanoparticles), P10, 20 and 30% (propolis extract at different concentrations). Fifteen treated strawberries with 3 repetitions were used and an Anova and tukey test ($p \leq 0.05$) were performed.

Chitosan and propolis have separately proven to be effective in increasing antioxidant capacity. However, in this research, a synergistic effect was also observed between CSNPs and P as the highest concentrations of phenols, flavonoids and % DPPH were obtained using these coatings. Currently, there is little information on the effect of combining these compounds on the antioxidant activity on fruit, although some studies have demonstrated the synergism of nanoparticles with other compounds in edible coatings. For example, Yang et al. (2016) [39] found that using lignin nanoparticles in polylactic acid-based films was highly efficient in terms of antioxidant capacity and, in combination with cellulose nanocrystals, a positive synergistic effect was generated in the antioxidant response of the films in vitro. In other work, Yang et al. (2016) [40] observed an increased antioxidant capacity through the addition of lignin nanoparticles to films based on polyvinyl alcohol-chitosan.

The synergism between the chitosan and the nanoparticles can be explained by the ability of the first compound to eliminate chelated ions and free radicals, thus avoiding hydrogen donation and resulting in greater antioxidant capacity. Their small size and low molecular weight means that the nanoparticles contribute to significant changes in the functional properties due to an increase in the surface area in relation to the volume. Therefore, they are more biologically active, improving the bioavailability of active ingredients and controlled release, and contributing to preserving the antioxidant capacity of the fruit [41,42].

With respect to the sensory evaluation, the coatings were evaluated satisfactorily and no statistical differences were observed between the treatments (Table 7). The coatings did not modify the taste or cause any bad odors. These data align with those of Marquez et al. (2009) [43] who evaluated a coating based on chitosan 0.6% and sucroester fatty acids (1%), and observed that its application to loquat fruit did not modify its flavor, aroma, or appearance.

Table 7. Sensory characteristics of strawberry fruit treated with different nanostructured coatings.

Coatings	Aroma	Color	Flavor
CS	7.2 ± 1.6 [a]*	8.3 ± 0.9 [a]	7.8 ± 1.4 [a]
CS+CSNP	7.1 ± 1.4 [a]	7.8 ± 1.1 [a]	7.4 ± 0.9 [a]
CS+CSNP+P10%	6.6 ± 1.5 [a]	7.2 ± 1.5 [a]	6.6 ± 1.8 [a]
CS+CSNP+P20%	7.0 ± 2.2 [a]	7.1 ± 1.9 [a]	7.2 ± 1.8 [a]
CS+CSNP+P30%	6.6 ± 1.7 [a]	7.2 ± 1.1 [a]	6.7 ± 1.8 [a]
Control	6.4 ± 1.5 [a]	7.8 ± 0.7 [a]	8.0 ± 1.3 [a]

* Means with equal letters (a) are not significantly different. ANOVA and Tukey test ($p \leq 0.05$). CS (chitosan), CSNPs (chitosan nanoparticles), P10, 20, and 30% (propolis extract at different concentrations). Ten strawberries were used per treatment and an Anova and tukey test ($p \leq 0.05$) were performed.

4. Conclusions

Normal ripening behavior was obtained in the coated fruit. In addition, strawberries coated with nanostructured chitosan and propolis extract, regardless of the concentration, yielded higher levels of the total phenols, flavonoids, and antioxidant capacity at the end of the 8th storage period compared with the untreated fruit. Furthermore, the application of the nanostructured coatings did not modify the sensory characteristics. The use of nanostructured chitosan coatings and propolis could be a viable alternative for preserving the quality and antioxidant capacity of strawberries.

Author Contributions: M.d.C.M.-G. performed the experiments; S.B.-B., planned and revised the experiments, and wrote the paper; Z.N.C.-P., advised and designed the nanocoatings, R.I.V.-A., advised and revised the biochemical and sensory tests; M.L.C.-R., performed the experiments; J.C.d.R.-G., obtained and prepared the propolis; M.d.L.R.-G., planned and revised the experiments and wrote the paper. All authors have read and agreed to the published version of the manuscript.

Funding: This research received no external contribution.

Conflicts of Interest: The authors declare no conflict of interest.

References

1. Galetto, C.D.; Verdini, A.R.; Zorrilla, S.E.; Rubiolo, C.A. Freezing of strawberries by immersion in CaCl2 solutions. *Food Chem.* **2010**, *123*, 243–248. [CrossRef]
2. Ruiz, M.; Ávila, J.; Ruales, J. Recubrimiento comestible bioactivo para aplicarlo en la frutilla (Fragaria vesca) como proceso de postcosecha. *Rev. Iberoam. Tecnol. Postcosecha* **2016**, *17*, 276–287.
3. Landanilla, M.S. Fruit morphology, anatomy and phisiology. In *Citrus Fruit. Biology, Technology and Evaluation*, 1st ed.; Ladanilla, M., Ed.; Academia Press: Icar, India, 2008; pp. 103–124.
4. Ghidelli, C.; Pérez-Gago, M.B. Recent advances in modified atmosphere packaging and edible coatings to maintain quality of fresh-cut fruits and vegetables. *Crit. Rev. Food Sci. Nutr.* **2018**, *58*, 662–679. [CrossRef] [PubMed]
5. Gol, N.B.; Ramana, R.T.V. Banana fruit ripening as influenced by edible coatings. *Int. J. Fruit Sci.* **2011**, *11*, 119–135. [CrossRef]
6. Bello-Lara, J.E.; Balois-Morales, R.; Juárez-López, P.; Alia-Tejacal, I.; Peña-Valdivia, C.B.; Jiménez-Zurita, J.O.; Sumaya-Martinez, T.; Jimenez-Ruiz, E. Coatings based on starch and pectin from 'Pear' banana (Musa ABB), and chitosan applied to postharvest 'Ataulfo' mango fruit. *Rev. Chapingo Ser. Hortic.* **2016**, *22*, 209–218. [CrossRef]
7. Hong, K.; Xie, J.; Zhang, L.; Sun, D.; Gong, D. Effects of chitosan coating on postharvest life and quality of guava (*Psidium guajava* L.) fruit during cold storage. *Sci. Hortic.* **2012**, *144*, 172–178. [CrossRef]
8. Gol, N.B.; Chaudhari, M.L.; Rao, T.V.R. Effect of edible coatings on quality and shelf life of carambola (Averrhoa carambola L.) fruit during storage. *J. Food Sci. Technol.* **2015**, *52*, 78–91. [CrossRef]

9. Baldoni, D.; Ventura-Aguilar, R.I.; Hernández-López, M.; Corona-Rangel, M.L.; Barrera-Necha, L.L.; Correa-Pacheco, Z.; Bautista-Baños, S. Calidad postcosecha de higos 'black mission' tratados con cubiertas naturales. *Rev. Iberoam. Tecnol. Postcosecha* **2016**, *17*, 267–275.
10. Araujo, G.P.; Souza, S.J.; Silva, M.V.; Yamashita, F.; Gonçalves, O.H.; Leimann, F.V.; Shirai, M.A. Physical, antimicrobial and antioxidant properties of starch-based film containing ethanolic propolis extract. *Int. J. Food Sci. Technol.* **2015**, *50*, 2080–2087. [CrossRef]
11. Zambrano-Zaragoza, M.L.E.; Mercado-Silva, E.; Gutiérrez-Cortez, E.; Castaño-Tostado, E.; Quintanar-Guerrero, D. Optimization of nanocapsules preparation by the emulsion–diffusion method for food applications. *LWT Food Sci. Technol.* **2011**, *44*, 1362–1368. [CrossRef]
12. Ortiz-Duarte, G.; Pérez-Cabrera, L.E.; Artés-Hernández, F.; Martínez-Hernández, G.B. Ag-chitosan nanocomposites in edible coatings affect the quality of fresh-cut melon. *Postharvest Biol. Technol.* **2019**, *147*, 174–184. [CrossRef]
13. Divya, K.; Smitha, V.; Jisha, M. Antifungal, antioxidant and cytotoxic activities of chitosan nanoparticles and its use as an edible coating on vegetables. *Int. J. Biol. Macromol.* **2018**, *114*, 572–577. [CrossRef]
14. Mohammadi, A.; Hashemi, M.; Hosseini, S. Chitosan nanoparticles loaded with Cinnamomum zeylanicum essential oil enhance the shelf life of cucumber during cold storage. *Postharvest Biol. Technol.* **2015**, *110*, 203–213. [CrossRef]
15. Eshghi, S.; Hashemi, M.; Mohammadi, A.; Badii, F.; Mohammadhoseini, Z.; Ahmadi, K. Effect of nanochitosan-based coating with and without copper loaded on physocochemical and bioactive components of fresh strawberry fruit (*Fragaria* × *ananassa* Duchesne) during storage. *Food Bioprocess Technol.* **2014**. [CrossRef]
16. Fernández, D.V.; Bautista-Baños, S.; Ocampo-Ramírez, A.; García-Pereira, A.; Falcón-Rodríguez, A. Películas y recubrimientos comestibles: Una alternativa favorable en la conservación poscosecha de frutas y hortalizas. *Rev. Cienc. Téc. Agropecu.* **2015**, *24*, 52–57.
17. Barrera, E.; Gil, M.; García, P.; Durango, D.; Gil, H. Empleo de un recubrimiento formulado para el manejo postcosecha de frutos de papaya (Carica papayal. Cv Hawaiana). *Rev. Fac. Nac. Agron.* **2012**, *65*, 6497–6506.
18. Ali, A.; Ling, W.; Zahid, N.; Kying, M. Efficacy of propolis and cinnamon oil coating in controlling post-harvest anthracnose and quality of chilli (Capsicum annum L) during cold storage. *Food Bioprocess Technol.* **2014**, *7*, 2742–2748. [CrossRef]
19. Siripatrawan, U.; Vitchayakitti, W. Improving functional properties of chitosan films as active food packaging by incorporating with propolis. *Food Hydrocoll.* **2016**, *61*, 695–702. [CrossRef]
20. Correa-Pacheco, Z.; Bautista-Baños, S.; Ramos-García, M.; Martinez-González, M.; Hernández-Romano, J. Physicochemical characterization and antimicrobial activity of edible propolis-chitosan nanoparticle films. *Prog. Org. Coat.* **2019**, *137*, 105326. [CrossRef]
21. Singleton, V.L.; Rossi, J.A. Colorimetry of total phenolics with phosphomolybdic-phosphotungstic acid reagents. *Am. J. Enol. Vitic.* **1965**, *16*, 44–58.
22. Chougui, N.; Tamendjari, A.; Hamidj, W.; Hallal, S.; Barras, A.; Richard, T.; Larbat, R. Oil composition and characterisation of phenolic compounds of Opuntia ficusindica seeds. *Food Chem.* **2013**, *139*, 796–803. [CrossRef] [PubMed]
23. Durazzo, A.; Lucarini, M. A current shot and re-thinking of antioxidant research strategy. *Braz. J. Anal. Chem.* **1995**, *20*, 9–11. [CrossRef]
24. Brand-Williams, W.; Cuvelier, M.; Berset, C. Use of a free radical method to evaluate antioxidant activity. *LWT Food Sci. Technol.* **1995**, *28*, 25–30. [CrossRef]
25. Gol, N.B.; Patel, P.R.; Rao, T.V.R. Improvement of quality and shelf-life of strawberries with edible coatings enriched with chitosan. *Postharvest Biol. Technol.* **2013**, *85*, 85–195. [CrossRef]
26. Maqbool, M.; Ali, A.; Alderson, P.G.; Zahid, N.; Siddiqui, Y. Effect of a noveledible composite coating based on gum arabic and chitosan on biochemical andphysiological responses of banana fruit during cold storage. *J. Agric. Food Chem.* **2011**, *59*, 5474–5482. [CrossRef]
27. Restrepo, J.; Aristizabal, T. Conservación de fresa (Fragaria x ananassa Duch cv. Camarosa) mediante la aplicación de recubrimientos comestibles de gel mucilaginoso, de penca sábila (Aloe barbadensis Miller) y cera de carnauba. *Rev. Univ. Quím. Farm.* **2010**, *17*, 252–263.

28. Ventura-Aguilar, R.I.; Bautista-Baños, S.; Flores-García, G.; Zavaleta-Avejar, L. Impact of chitosan based edible coatings functionalized with natural compounds on Colletotrichum fragariae development and the quality of strawberries. *Food Chem.* **2018**, *262*, 142–149. [CrossRef]
29. Pilon, L.; Spricigo, P.C.; Miranda, M.; de Moura, M.R.; Assis, O.B.G.; Mattoso, L.H.C.; Ferreira, M.D. Chitosan nanoparticle coatings reduce microbial growth on fresh-cut apples while not affecting quality attributes. *Int. J. Food Sci. Technol.* **2014**, *50*, 440–448. [CrossRef]
30. Bodini, R.; Sobral, P.; Favaro-Trindade, C.; Carvalho, R. Properties of gelatin-based films with added ethanol–propolis extract. *LWT Food Sci. Technol.* **2013**, *51*, 104–110. [CrossRef]
31. Pastor, C.; Sánchez-González, L.; Cháfer, M.; Chiralt, A.; González-Martínez, C. Physical and antifungal properties of hydroxypropylmethylcellulose based films containing propolis as affected by moisture content. *Carbohydr. Polym.* **2010**, *82*, 1174–1183. [CrossRef]
32. Trejo, M.; Ramos, K.; Pérez, C. Efecto de la aplicación de un recubrimiento comestible a base de gelatina sobre la calidad de fresa (Fragaria vesca L.) almacenada en refrigeración. In *En V Congreso Iberoamericano de Tecnología Postcosecha y Agroexportaciones*; (pp. 230–239); Asociación Iberoamericana de Tecnología Postcosecha (AITEP): Murcia, Spain, 2007; pp. 236–237.
33. Del-Valle, V.; Hernández-Muñoz, P.; Guarda, A.; Galloto, M.J. Development of a cactus-mucilage edible coating (Opuntia ficus Indica) and its application to extend strawberry shelf-life. *Food Chem.* **2005**, *91*, 751–756. [CrossRef]
34. Wang, S.Y.; Gao, H. Effect of chitosan-based edible coating on antioxidants, antioxidant enzyme system, and postharvest fruit quality of strawberries (*Fragaria ananassa* Duch.). *LWT Food Sci. Technol.* **2013**, *52*, 71–79. [CrossRef]
35. López-Mata, M.; Ruiz-Cruz, S.; Navarro-Preciado, C.; Ornelas-Paz, J.; Estrada-Alvarado, M.; Gassos-Ortega, L.; Rodrigo-García, J. Efecto de recubrimientos comestibles de quitosano en la reducción microbiana y conservación de la calidad de fresas. *Rev. Cienc. Biol. Salud* **2012**, *14*, 33–43. [CrossRef]
36. Zahid, N. Efficacy of ethanolic extract of propolis in maintaining postharvest quality of dragon fruit during storage. *Postharvest Biol. Technol.* **2013**, *79*, 69–72. [CrossRef]
37. Thomas, A.; Resende, R.C.; Vilas, A.C.; Olivera, L. Cassava starch edible coating incorporated with propolis on bioactive compounds in strawberries. *Ciênc. Agrotec.* **2016**, *40*, 87–96. [CrossRef]
38. Anjum, S.I.; Ullah, A.; Khan, K.A.; Attaullah, M.; Khan, H.; Ali, H.; Bashir, M.; Ansari, M.; Ghramh, H.; Adgaba, N.; et al. Composition and functional properties of propolis (bee glue): A review. *Saudi J. Biol. Sci.* **2018**, *26*, 1695–1703. [CrossRef]
39. Yang, W.; Fortunati, E.; Dominici, F.; Giovanale, G.; Mazzaglia, A.; Balestra, G.M.; Kenny, J.M.; Puglia, D. Effect of cellulose and lignin on disintegration, antimicrobial and antioxidant properties of PLA active films. *Int. J. Biol. Macromol.* **2016**, *89*, 360–368. [CrossRef] [PubMed]
40. Yang, W.; Owczarek, J.S.; Fortunati, E.; Kozanecki, M.; Mazzaglia, A.; Balestra, G.M.; Kenny, J.M.; Torre, L.; Puglia, D. Antioxidant and antibacterial lignin nanoparticles in polyvinyl alcohol/chitosan films for active packaging. *Ind. Crops Prod.* **2016**, *94*, 800–811. [CrossRef]
41. Kerch, G. The potential of chitosan and its derivatives in prevention and treatment of age-Related Diseases. *Mar. Drugs* **2015**, *13*, 2158–2182. [CrossRef] [PubMed]
42. Radhakrishnan, Y.; Gopal, G.; Lakshmanan, C.C.; Nandakumar, K.S. Chitosan nanoparticles for generating novel systems for better applications: A review. *J. Mol. Genet. Med.* **2015**, *S4*, 5.
43. Marquez, C.; Cartagena, J.; Pérez-Gago, M. Efecto de recubrimientos comestibles sobre la calidad en postcosecha del níspero japonés (*Eriobotrya japónica* T.). *Rev. Fac. Quím. Farm.* **2009**, *16*, 304–310.

coatings

MDPI

Article

Individual and Combined Coatings of Chitosan and Carnauba Wax with Oregano Essential Oil to Avoid Water Loss and Microbial Decay of Fresh Cucumber

María Melissa Gutiérrez-Pacheco [1], Luis Alberto Ortega-Ramírez [1], Brenda Adriana Silva-Espinoza [1], Manuel Reynaldo Cruz-Valenzuela [1], Gustavo Adolfo González-Aguilar [1], Jaime Lizardi-Mendoza [1], Raquel Miranda [2] and Jesús Fernando Ayala-Zavala [1,*]

1 Centro de Investigación en Alimentación y Desarrollo, A.C., Carretera Gustavo Enrique Astiazarán Rosas, No. 46, Col. La Victoria, Hermosillo, Sonora 83304, Mexico; melissa.gtzpacheco@gmail.com (M.M.G.-P.); ingeluis_100@hotmail.com (L.A.O.-R.); bsilva@ciad.mx (B.A.S.-E.); reynaldo@ciad.mx (M.R.C.-V.); gustavo@ciad.mx (G.A.G.-A.); jalim@ciad.mx (J.L.-M.)

2 Department of Biochemistry and Molecular Biology, Federal University of Ceara, Av. Mister Hull 2297, Fortaleza, Ceara 60440-554, Brazil; rmiranda@ufc.br

* Correspondence: jayala@ciad.mx

Received: 20 May 2020; Accepted: 26 June 2020; Published: 29 June 2020

Abstract: The objective of the present study is to evaluate the effect of individual and combined coatings of chitosan (0.008 g·mL^{-1}) and carnauba wax (0.1 g·mL^{-1}) with oregano essential oil (OEO, 0.08 g·mL^{-1}) to reduce dehydration and microbial decay of fresh cucumbers stored at 10 °C. Chitosan-OEO-wax films showed the lowest water vapor transmission rate (0.141 g·m^{-2}·h^{-1}), compared to single chitosan films (0.257 g·m^{-2}·h^{-1}). While chitosan-OEO films completely inhibited the in vitro growth of *Alternaria alternata* and reduced the growth of *Salmonella* Typhimurium, *Escherichia coli* O157:H7, mesophilic bacteria, and fungi isolated from decayed cucumbers. Besides, the infrared analysis of chitosan-OEO-wax films showed shifts in O–H and N–H absorption bands, indicating possible hydrogen bonding between the components. Wax and wax-OEO were the most effective coatings to prevent weight loss in cucumbers during 15 days of storage at 10 °C, while the most effective antimicrobial treatments were chitosan and chitosan-OEO. Therefore, these results showed that carnauba wax and carnauba wax-OEO coatings were the most effective in weight loss, whereas chitosan and chitosan-OEO were the most effective to reduce the microbial load of the treated fresh cucumber.

Keywords: *Cucumis sativus* L.; dehydration; antimicrobial activity; bilayer coating; essential oils

1. Introduction

Cucumber (*Cucumis sativus* L.) is a low-calorie fruit belonging to the *Cucurbitaceae* family, non-climacteric, harvested, and consumed at an immature stage. It is a rich source of potassium, magnesium, iron, with a high water content—approximately 90% [1]. The quality of cucumber is based on its shape uniformity, dark green color, firmness, size, absence of defects, and rot [2]. However, during post-harvest, the cucumber is highly susceptible to physiological changes, microbial decay, and dehydration with the consequent loss of appearance, and nutritional quality [3]. Also, it is susceptible to the attack of bacteria (e.g., *Erwinia* spp. and *Xanthomonas* spp.) and fungi (e.g., *Alternaria* spp. and *Rhizopus* spp.) [4]. Commonly, some of these problems are solved with the use of low storage temperatures; however, cucumbers are sensitive to chilling injury below 10 °C; for this reason, they are stored at 10–12.5 °C, becoming more susceptible to quality loss after 14 days. These problems justify the need for developing adequate water loss barrier and antimicrobial techniques to preserve the cucumbers' quality.

Edible coatings can act as barriers to reduce water loss and gas exchange, depending on their composition [5]. Different materials are used to formulate edible coatings, including proteins, lipids, and polysaccharides. Polysaccharides have excellent gas barrier properties; however, they are highly hydrophilic and show high water vapor permeability. On the other hand, lipids are hydrophobic compounds with water barrier properties, although, their nonpolymeric nature limits their ability to form films with good mechanical integrity [5]. Among polysaccharides, chitosan obtained from alkaline deacetylation of chitin has been widely used because of its antimicrobial properties [6]. Chiabrando et al. [7], reported that chitosan coatings significantly reduced microbial decay of minimally processed nectarines compared to control fruits. Similarly, chitosan coatings (2%) in broccoli florets stored at 5 °C resulted in a significant reduction of total mesophilic and psychrotrophic bacteria; also, a bactericidal effect was observed in the inoculated *E. coli* O157:H7 [8]. In addition, Pavinatto et al. [9] reported a reduction of gray fungus attack and an insignificant alteration in flavor, appearance, aroma, and texture of strawberries coated with 1% chitosan and stored for 7 days at 25 °C. Other polysaccharide-based coatings also have been applied to improve the quality parameters of fruits. *Aloe vera* coatings (3%) improved firmness and total soluble solids of tomato; however, no effect was observed in weight loss compared to control [10]. Similarly, alginate coatings showed a good gas permeability and maintained the firmness of fresh-cut melon; however, no antimicrobial activity and weight loss protection were observed. Besides their different benefits, polysaccharides-based coatings are considered a low water barrier [11].

On the contrary, natural waxes (carnauba, shellac, and beeswax) limit the water loss better than polysaccharides-based coatings. Carnauba wax is a lipid-based material obtained from *Copernicia cerifera* leaves, predominantly comprised of aliphatic esters and diesters of cinnamic acid with a high melting point and low solubility [12]. These characteristics grant the relatively inert and stable character to carnauba wax; in addition, it is generally recognized as safe by the Food and Drug Administration of the United States and normally used in edible coatings formulations [5,12]. Won and Min [13] reported a significant reduction in the weight loss of Satsuma mandarins treated with carnauba wax (18.1%) during storage at 4 and 25 °C for 28 days compared with control fruits. In addition, the weight loss was significantly reduced in Valencia oranges and avocados after the application of carnauba wax [14,15]. However, it is important to note that these authors did not evaluate the antimicrobial effect of these coatings, microbial decay being one of the main factors that compromise the postharvest life of these fruits, highlighting the need for an additional coating material as our study proposes. Therefore, it can be expected that a combination of constituents could obtain better results.

Edible coatings may carry antimicrobial additives protecting fresh produce of postharvest decay [16]. Oregano essential oil (OEO) possess antimicrobial activity, and it may be added to edible coatings as a food additive; its efficacy has been proved when added into several edible coatings [17,18]. The antibacterial activity of OEO is attributed to the constituents carvacrol and thymol; as a result of their lipophilic nature, these monoterpenes can become embedded in the bacteria or fungi membrane causing the alteration of enzymatic systems, cellular disruption, and loss of cellular constituents [19,20]. Some studies reported the efficacy of the combination of chitosan and carnauba wax with essential oils to maintain the postharvest quality of cucumber and other fruits. Mohammadi et al. [21] reported that nanochitosan-based coating loaded with *Zataria multiflora* essential oil improved physicochemical quality and significantly reduced total bacterial, yeasts, and molds counts of cucumber stored at 10 °C for 21 days. Won and Min [13] also reported that the addition of OEO provides an antimicrobial activity to carnauba wax coatings. However, these studies did not characterize the physicochemical properties of the coatings, which is important to compare the responses with those observed in vivo. Therefore, the present study evaluated the effect of individual and combined coatings of chitosan and carnauba wax with OEO to reduce dehydration and microbial decay of fresh cucumbers.

2. Materials and Methods

2.1. Chemicals

Medium molecular weight chitosan (75%–85% deacetylation), carnauba wax yellow No.1, glycerol, and anhydrous calcium chloride were obtained from Sigma-Aldrich (St. Louis, MO, USA). Glacial acetic acid was obtained from JT Baker (Madrid, Spain), while the coconut oil was obtained from Oils by Nature Inc. (Solon, OH, USA).

2.2. Plant Materials

Cucumber (*Cucumis sativus* L.) fruits were harvested in a field in Hermosillo, Sonora, Mexico (29°17′20.4″ N 110°54′35.0″ W) after 57 days from flowering, at a slightly immature state, with a dark green color visually determined by using the color comparator for medium green CC-1 proposed by the USDA, uniform in shape, size, weight and free from growth defects and decay; in addition, firmness, CO_2 production and color of cucumbers were determined (Table 1). OEO (*Lippia graveolens* L.) was obtained from ORE Procesadora de Oregano Silvestre [22], Chihuahua, Mexico.

Table 1. Firmness, CO_2 production, and color of cucumbers.

Parameter	Mean ± Standard Deviation
Firmness (N)	48.80 ± 1.89
CO_2 production rate (mL CO_2/kg·h)	1.10 ± 0.65
Color	–
Lightness	29.21 ± 1.01
Chroma	59.50 ± 13.19
Hue	132.76 ± 0.54

2.3. Film Formulation

For chitosan films, 0.8 g of chitosan and 0.2 g of glycerol were dispersed in 100 mL of 1% glacial acetic acid solution and stirred at 25 °C for 24 h. The filmogenic mixture was filtered using cheesecloth to remove small impurities, sterilized at 121 °C for 15 min, and cooled at 25 °C. Subsequently, 20 mL of the chitosan dispersion was cast on Petri dishes with diameters of 8 cm and dried at 25 °C for 24 h. Then the films were peeled off from the plates and stored in a desiccator for further analysis. For carnauba wax films, a 10% carnauba wax solution was made by melting 10 g of wax in 90 mL of coconut oil heated at 70 °C with constant stirring. Bilayer films (chitosan-carnauba wax and chitosan-OEO-carnauba wax) were elaborated by brushing 0.5 mL of the carnauba wax on the preformed chitosan films. For the films added with OEO, 0.2 g of Tween 20 and 8 mg·mL^{-1} of OEO (this concentration was selected based on the minimal inhibitory concentration of OEO against microbiota of decayed cucumbers) were added directly into the chitosan and carnauba wax for their individual coatings, and in the chitosan-wax coating the oil was added into the carnauba wax. Each formulation was mixed at 13,500 rpm for 5 min in a Kinematica Polytron homogenizer PT 1200C (Cambridge Scientific Products, Watertown, MA, USA).

2.4. Characterization of the Formulated Films

2.4.1. Thickness and Water Vapor Transmission (WVT)

Film thickness was measured with a digital micrometer (E.J. Cady and Co., Wheeling, IL, USA); three measurements were carried out at different points of 5 films per treatment (chitosan, chitosan-wax, chitosan-OEO, and chitosan-OEO-wax) and results were expressed in millimeters (mm). WVT was determined gravimetrically based on the American Society for Testing and Materials method (ASTM) [23]; for this, each film was placed in the top of a moisture permeation cell (118.64 cm^3 volume) with 30 g of dried calcium chloride beads to ensure a relative humidity (RH) of 0% inside the cell. Subsequently, the cells were placed in a desiccator (volume 1500 cm^3) containing 70 mL of a saturated solution of magnesium nitrate to ensure 90% RH and left at 25 °C for 8 h. The water mass transferred through the film and adsorbed by the calcium chloride was determined by the weight gained for each permeation cell. Cells were weighed each hour for eight hours, and the slope of weight gain vs. time was obtained by linear regression. The following formula determined the WVT of the films: WVT = m/a, where m is the slope of weight gain vs. time, and a is the film area where the mass transfer occurred. The analysis was performed by triplicate, and results were expressed as grams of water per square meter per hour (g $m^{-2}{\cdot}h^{-1}$). Because of the experimental complexity, the thickness and WVTR of the individual wax films were not evaluated.

2.4.2. In Vitro Antimicrobial Capacity of the Formulated Films

Cucumber microbiota (mesophilic bacteria and fungi) isolated from decayed fruits, *Salmonella enterica* subspecies *enterica* serovar Typhimurium (ATCC® 14028), *Escherichia coli* O157:H7 (ATCC® 43890), and *Alternaria alternata* (ATCC® 6663) were exposed to the formulated films. The antimicrobial assay was carried out based on the dilution method described by the National Committee for Clinical Laboratory Standards [24], with some modifications. For the antimicrobial assay, 20 mg of each film (2 mg·mL^{-1}) was introduced into tubes containing 10 mL of Mueller Hinton or potato dextrose broth for bacteria and fungi, respectively. After 15 min, tubes were inoculated with 1×10^6 colony-forming units per milliliter (CFU mL^{-1}) of each challenged bacteria or fungi. Subsequently, 1.0 mL from each tube was plated in Mueller Hinton or acidified potato dextrose agar and incubated at 37 ± 2 °C for 24 h and 25 ± 2 °C for five days for bacteria and fungi, respectively. The analysis was performed by triplicate, the microbial colonies were counted, and the results were expressed as Log CFU mL^{-1} and compared with a control without films.

2.4.3. Fourier Transform Infrared Spectroscopy (FTIR)

FTIR spectra (Instrument Corp. Madison, WI, USA) were obtained to evaluate physicochemical interactions among the film components (chitosan, wax, chitosan-wax, chitosan-OEO, and chitosan-OEO-wax). Data were recorded in the transmission mode using a spectrophotometer FTIR Nicolet Protegé 460 (Instrument Corp. Madison, WI, USA) under a wavenumber range of 4000–400 cm^{-1} with a resolution of 4 (cm^{-1}), taking 64 scans at a rate of 0.63 (s^{-1}). Solid samples were pressed within KBr pellets, and liquid samples (OEO and carnauba wax) were placed over preformed pellets, respectively.

2.5. *Effect of the Formulated Coatings on Water Loss and Microbial Decay of Fresh Cucumbers*

2.5.1. Coating Application

A total of 252 cucumbers (whole and unpeeled) were used in this experiment; 216 fruits were divided into six groups and coated with each treatment, and 36 fruits were left uncoated as controls. Cucumbers were washed with chlorinated water (200 ppm) for 3 min and air-dried at 25 °C, and coated manually with brushes (1.5-inch-wide, Maxtool®, Mexico City, Mexico), adding uniformly 0.5 mL of each treatment per cucumber. For applying bilayer coatings, cucumbers were first coated with chitosan and dried with forced air at 25 °C for 10 min; afterward, wax (with and without OEO) was applied, then cucumbers were stored on polypropylene trays at 10 °C and 90% RH. The effects of edible coatings on the weight loss and microbial spoilage of the fruit were assessed in three trays with two cucumbers per treatment at 0, 3, 6, 9, 12 and 15 days.

2.5.2. Fruit Weight Loss

The fruit was weighed at 0, 3, 6, 9, 12 and 15 days of storage at 10 °C, and the slope of weight loss vs. time was obtained by linear regression. Weight loss was measured using the equation: $(A - B)/A$, where A was the initial weight (g) at t_0, while B was the weight (g) at a given storage time (t_n). The analysis was performed by triplicate, and results were expressed as weight loss (%).

2.5.3. Changes of Microbial Load of the Coated Fruit

Total mesophilic bacteria, molds and yeasts were counted on 0, 3, 6, 9, 12 and 15 days of storage at 10 °C. Cucumbers were sampled (10 g) and homogenized for 1 min in 90 mL of peptone water. Subsequently, decimal dilutions were made, and 1 mL of each sample was poured on plate count agar or potato dextrose acidified agar, and incubated at 37 ± 2 °C for 24 h or 25 ± 2 °C for five days, for mesophilic bacteria and for molds and yeasts, respectively [25]. The analysis was performed with four replicates, and results were expressed as Log CFU·g^{-1}.

2.6. Statistical Analysis

The effect of the films composition (chitosan, chitosan-wax, chitosan-OEO, wax-OEO, and chitosan-OEO-wax) on the WVT, thickness, and in vitro antimicrobial capacity was evaluated with a completely randomized design. The effect of the edible coatings over dehydration and microbial spoilage was a completely randomized experimental design with a factorial arrangement (5 × 6), where the factors were the coating treatments (chitosan, wax, chitosan-wax, chitosan-OEO, wax-OEO, and chitosan-OEO-wax) and the storage time (0, 3, 6, 9, 12, 15 days), and the analyzed responses were weight loss and microbial growth. All experiments were done by triplicate, expressing the results as means ± standard deviation. An analysis of variance (ANOVA) was performed, and means were compared by the Tukey Kramer test ($p \leq 0.05$) using the statistical software NCSS 2007 (NCSS, LLC, Kaysville, UT, USA).

3. Results

3.1. Characterization of the Formulated Films

3.1.1. FTIR Spectra

Figure 1 shows the infrared spectra of OEO (a), chitosan (b), wax (c), chitosan-OEO (d), and wax-OEO (e) in the wavelength range of 4000–400 cm^{-1}. OEO spectrum showed at 3361 and 2958 cm^{-1} the characteristic peaks of O–H and C–H stretching, respectively; these signals could be attributed to the presence of carvacrol and thymol. On the other hand, the peaks of the C=O stretch (amide I) and flexion for the N–H group (amide II) of the chitosan were identified at a wavenumber around 1645 and 1554 cm^{-1}, respectively (Figure 1b). These peaks represent the N-acetylglucosamine structure, which could be found in chitosan with a low acetylation degree [26]. Also, at 3334 cm^{-1}, the corresponding absorption band of the O–H stretching was observed. On the other hand, a shift in the O–H stretching peak of the chitosan was observed ($\Delta_{OH} = 11$) when combined with OEO (Figure 1c), which could be attributed to the interactions between the terpenoids and chitosan. Figure 1d showed the spectrum of carnauba wax with the C–H and C=O stretching signals observed at 2930 and 1743 cm^{-1}, respectively. In contrast, no changes were observed in the wax spectrum after the addition of OEO, which indicated the lack of interaction between these components (Figure 1e).

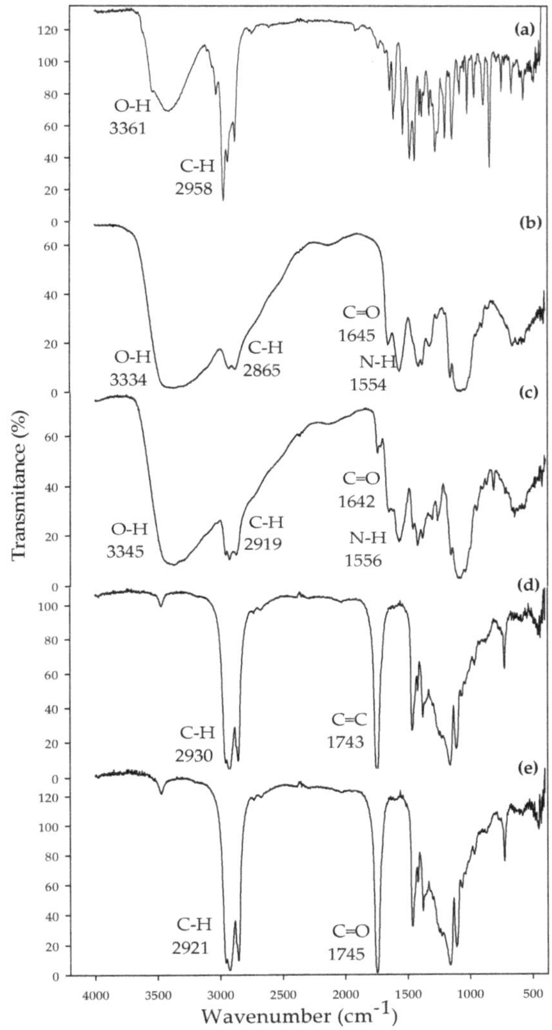

Figure 1. Fourier transform infrared spectroscopy (FTIR) spectra of oregano essential oil (OEO) (**a**), chitosan (**b**), chitosan-OEO (**c**), wax (**d**), and wax-OEO (**e**) films.

Moreover, the addition of wax on the chitosan film (chitosan-wax film) caused a shift toward higher frequencies in the absorption bands of the O–H (Δ_{OH} = 50) and C=O stretch of amide I (Δ_{CO} = 48.2), and the presence of the characteristic C=O peak of the fatty acid esters (Figure 2a). The same behavior was observed in the chitosan-OEO-wax film (Figure 2b), where the maxima absorption of the O–H (stretching), C=O (stretching), and N–H (flexion) peaks were shifted toward higher frequencies (Δ_{OH} = 7.7, Δ_{CO} = 44.85, and Δ_{NH} = 14). This behavior reflected that the incorporation of wax or OEO on the chitosan films could affect the maxima absorption peaks of the groups responsible for the antibacterial activity. Similar results were reported by Kaya et al. [27], who observed a shift of the N–H and O–H absorption peaks of chitosan after the addition of *Berberis crataegina* fruit extract, indicating hydrogen bond formation between the NH_2 group of the chitosan molecule and the O–H group of the fruit phenolic compounds. It is well-known that changes in the absorption spectra (in frequency and shape) can be correlated with physicochemical interactions among the components [27]. In this sense, special attention must be given to O–H and N–H groups, considering that their electronic changes could cause variations in the properties of the film [28]. Based on this, an additional NMR analysis could be made to confirm the interactions between these components. Also, it is well-known that the antimicrobial activity of the OEO is related to the active hydroxyl groups of carvacrol and thymol [29]; therefore, any interaction in these sites could lead to a reduction of this property.

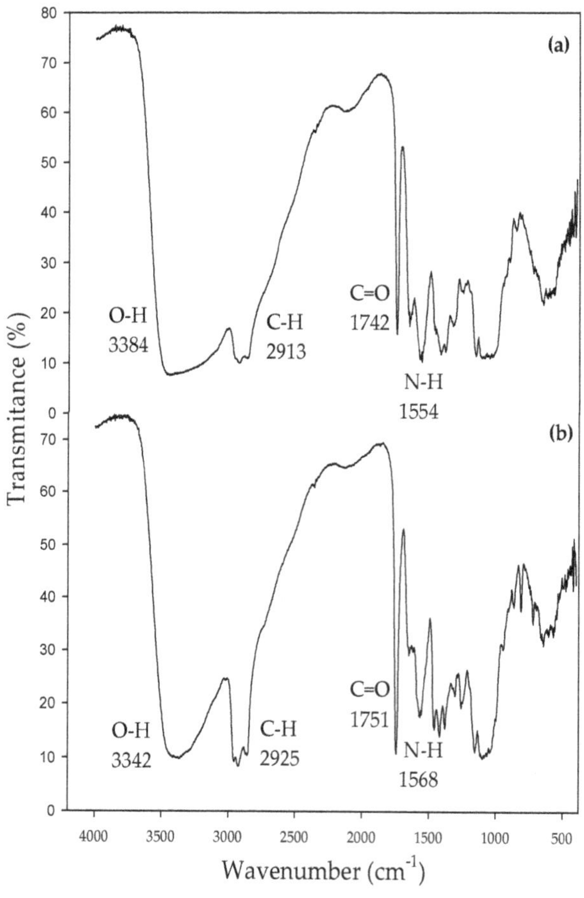

Figure 2. FTIR spectra of chitosan-wax (**a**) and chitosan-OEO-wax (**b**) films.

Several studies have tested the effect of the formulation on the film properties [30,31]; however, few studies have analyzed the physicochemical interactions among the constituents as the present study did. Considering that the presence and availability of certain functional groups grant the specific properties to the formulated materials, it becomes crucial to study their interactions and impact on the films functionality.

3.1.2. Thickness and WVT

Table 2 shows the results of thickness and WVT of the formulated films; the thickness was similar ($p \geq 0.05$) for chitosan, chitosan-wax, and chitosan-OEO, being the thickness of the chitosan-OEO-wax film the highest ($p \leq 0.05$) of all treatments. Chitosan films showed the highest WVT, followed by chitosan-OEO, with no significant differences between them ($p < 0.05$). Moreover, chitosan-wax and chitosan-OEO-wax films reduced the WVT by 35% and 45%, respectively, compared to chitosan films. The WVT and FTIR spectra results could support the discussed interaction of chitosan O–H groups with the wax functional groups (Figure 2a,b). In this sense, the decrease in WVT after the wax incorporation could be attributed to the limited availability of free O–H groups, which could cause a decrease in the absorption of water molecules and thus lower its permeability.

Table 2. Thickness and WVT of edible films constituted by chitosan, OEO, and carnauba wax at 25 °C.

Film	Thickness (mm) *	WVT (g/m^2 h) **
Chitosan	0.025 ± 0.007 ^{a***}	0.257 ± 0.013 ^{a}
Chitosan-OEO	0.022 ± 0.002 ^{a}	0.241 ± 0.019 ^{a}
Chitosan-Wax	0.027 ± 0.005 ^{a}	0.167 ± 0.012 ^{b}
Chitosan-OEO-Wax	0.037 ± 0.004 ^{b}	0.141 ± 0.008 ^{b}

* n = 15; ** n = 3; *** Different letters in the same column indicate significant differences ($p \leq 0.05$).

Several studies have shown a decrease in the water vapor permeability of polysaccharide films with the incorporation of lipidic compounds [30]. Ochoa, Almendárez, Reyes, Pastrana, López, Belloso and Regalado-González [30] developed composite films with low WVT and antimicrobial activity based on corn starch, beeswax, lauric alginate, and natamycin. However, they did not study the contribution of each component to the efficacy of the whole film, as the present study did. In the same study, an increment of thickness from 0.050 to 0.064 mm after beeswax incorporation was recorded. Similarly, Santos et al. [32] observed an improvement in the water barrier properties of chitosan films after the addition of beeswax. Haq et al. [33] reported a 50% lower water vapor permeability of gum Cordia films added with beeswax. Agar/maltodextrin films showed better barrier properties when beeswax was added as a bilayer [34]. Therefore, hydrophilic-based films with a coat of hydrophobic material could diminish the WVT. This lower WVT could be a good alternative to decreased weight loss in fresh produce caused by dehydration; this approach is shown in Section 3.2.1.

3.1.3. In Vitro Antimicrobial Activity of Films

Figure 3 shows the antimicrobial activity of edible films against bacteria (a) and fungi (b) isolated from the decayed cucumber. Figure 3a showed that after incubation at 37 °C for 24 h, chitosan-OEO films showed the greatest in vitro reduction ($p \leq 0.05$) of bacteria isolated from cucumber (3.23 Log CFU·mL^{-1}), while the wax-OEO films reduced by about 1 Log CFU·mL^{-1}. Also, chitosan and chitosan-wax films showed a reduction ($p \leq 0.05$) of 0.4–0.6 Log CFU·mL^{-1}; however, chitosan, wax, chitosan-wax, and chitosan-OEO-wax showed bacterial counts similar to the control with no differences among them ($p \geq 0.05$). A similar pattern can be observed in Figure 3b, where chitosan-OEO showed the highest reduction (1.12 Log CFU·mL^{-1}) of fungal counts compared to the control ($p \leq 0.05$). In addition, chitosan, wax, chitosan-wax, and chitosan-OEO-wax showed no differences among them ($p \geq 0.05$) against fungal counts; it has to be mentioned that these treatments showed fungal counts significantly lower than the control (0.47–0.63 Log CFU·mL^{-1} reduction).

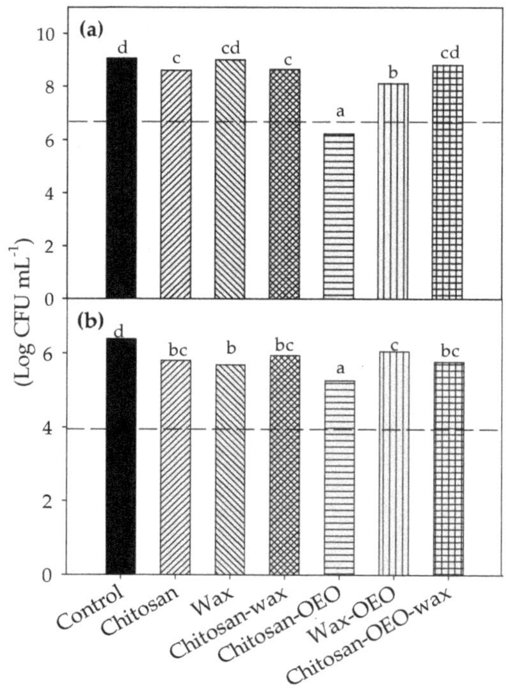

Figure 3. Antimicrobial activity of chitosan, carnauba wax, and oregano essential oil (OEO) films against mesophilic bacteria (**a**) and fungi (**b**) isolated from the decayed cucumber. Dotted lines signify the added inocula, and different letters among bars indicate significant differences ($p \leq 0.05$).

Figure 4 shows the in vitro antibacterial effect of edible films against *S.* Typhimurium (a) and *E. coli* O157:H7 (b). It was observed that the addition of OEO into chitosan films improved ($p \leq 0.05$) the antibacterial activity against *S.* Typhimurium (3.12 Log CFU·mL^{-1} reduction) compared to the control after 24 h of incubation at 37 °C (Figure 4a). On the other hand, chitosan and wax-OEO films significantly reduced *Salmonella* counts by 1.25 and 0.79 Log CFU·mL^{-1}, respectively. However, wax, chitosan-wax, and chitosan-OEO-wax films did not inhibit the growth of *S.* Typhimurium, showing counts similar ($p \geq 0.05$) to the control. On the other hand, for *E. coli* O157:H7, chitosan-OEO films reduced 3.45 Log CFU·mL^{-1} (Figure 4b), while chitosan and chitosan-wax reduced 1.58 and 0.43 Log CFU·mL^{-1}, respectively. Wax, wax-OEO, and chitosan-OEO-wax films showed no effect against *E. coli* O157:H7.

Figure 5 shows the fungicidal effect of the chitosan-OEO film against *A. alternata*. On the other hand, it was observed that chitosan-OEO-wax reduced 3 Log CFU·mL^{-1}, compared to the untreated fungus after five days at 25 °C. On the other hand, no differences were observed among chitosan, wax-OEO, and control, showing similar fungal counts. Furthermore, wax films did not inhibit the growth, showing counts 2.3–5 Log CFU·mL^{-1} higher ($p \leq 0.05$) than other treatments and control. The antimicrobial potential of chitosan-OEO films could be attributed to the action of each component. The antimicrobial activity of chitosan could be exerted by the interaction of its amino groups with the phospholipids in the bacterial membrane, leading to loss of functionality [28]. On the other hand, different studies suggest that carvacrol and thymol, major OEO components, may disintegrate the outer membranes of microorganisms because of their physicochemical interactions with lipids and proteins, causing a release of the cellular content and affecting viability [20]. It is important to mention that the reduced activity of wax-OEO and chitosan-OEO-wax films (versus the effect observed in chitosan-OEO films) could be attributed to a low diffusion of OEO compounds due to the different components of coating's structure or their affinity to the hydrophobic phase of wax, reducing the oils migration from the film. In this sense, more studies are needed to characterize the diffusion of OEO compounds throughout these coating systems.

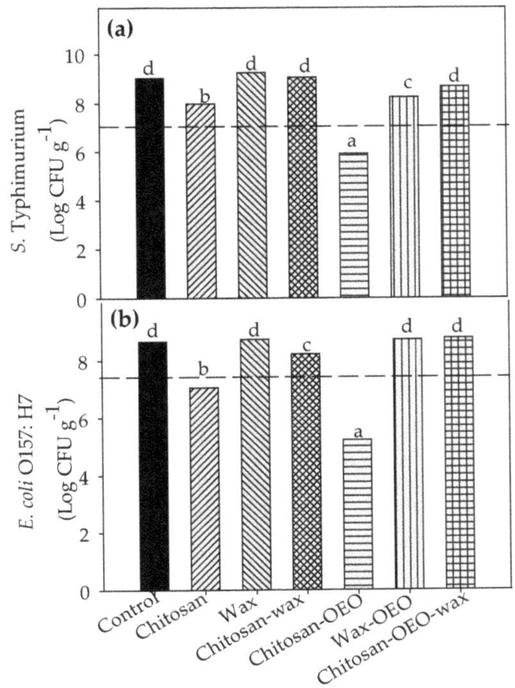

Figure 4. Antibacterial activity of chitosan, oregano essential oil (OEO), and carnauba wax films against *S.* Typhimurium (**a**) and *E. coli* O157:H7 (**b**) Dotted lines indicate the initial bacterial load, and different letters among bars indicate significant differences ($p \leq 0.05$).

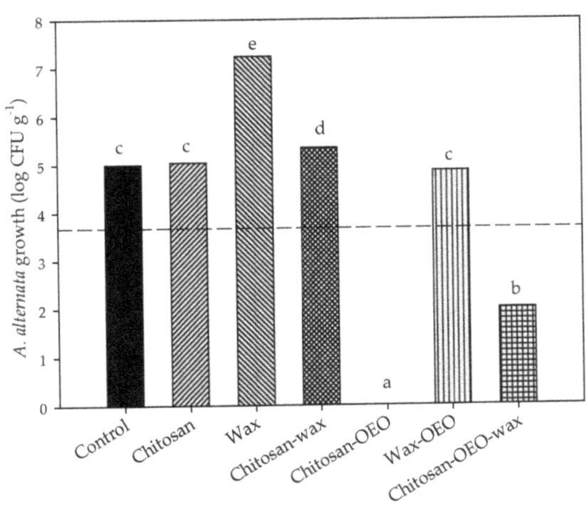

Figure 5. Antifungal activity of chitosan, oregano essential oil (OEO), and carnauba wax films against *A. alternata*. The dotted line indicates the initial fungal load, and the different letters indicate significant differences ($p \leq 0.05$).

Several studies have shown the ability of chitosan in carrying plant compounds to inhibit the growth of pathogenic bacteria [35]. Fernández-Pan et al. [36] reported that chitosan films with 20% carvacrol (the main terpene of OEO) reduced the growth of *Pseudomonas fragi, Shewanella putrefasciens,* and *Aeromonas hydrophila*. Similarly, Yuan et al. [37] observed an increase in the antibacterial activity of chitosan films against *Staphylococcus aureus* and *E. coli* after the addition of 10 mg·mL^{-1} of carvacrol. On the other hand, chitosan films incorporated with *Thymus piperella* essential oil reduced the growth of *Serratia marcescens* and *Listeria innocua* [38]. Chitosan alone or combined with OEO has also shown fungicidal activity against *Botrytis cinerea, Penicillium sp., Rhizopus stolonifer*, and *A. alternata*. As mentioned above, the polycationic structure of this compound is responsible for its antifungal properties. Chitosan could induce morphological and structural changes in fungal cells by causing molecular disorganization [28]. Finally, the incorporation of OEO within chitosan films improved their antimicrobial activity, which was reflected by the significant reduction of bacterial and fungal growth.

3.2. Postharvest Changes of Coated Cucumbers

3.2.1. Coating Influence on Postharvest Weight Loss

Table 3. shows the weight loss of coated cucumbers stored at 10 °C for 15 days. Wax and wax-OEO treatments significantly reduced cucumber weight loss by 0.18%–1.63% compared to control fruits, with wax coatings the most effective ($p \leq 0.05$) compared to the other treatments, which showed similar losses to those of control fruits. On the other hand, chitosan coatings caused no decrement of cucumber weight loss, which was expected given their hydrophilic characteristics. Based on the WVT results described in Section 3.1.2, it was expected that there would be a better barrier property when carnauba wax was added to chitosan film; however, no differences ($p \geq 0.05$) were observed between chitosan, chitosan-OEO, chitosan-wax, wax-OEO, and chitosan-OEO-wax. WVT was measured at 25 °C, while cucumbers were stored at 10 °C. This condition was corroborated by microphotographs from fruit coated with wax and stored at 25 and 10 °C. The carnauba wax coated cucumber stored at 25 °C showed a smooth and homogenous surface (Figure 6a), but the fruit with the same treatment and stored at 10 °C showed a rough surface (Figure 6b), with the appearance of crystals (Figure 6c). Meanwhile, the chitosan and chitosan-OEO coated cucumbers showed a homogeneous coated surface at both temperatures (data not shown). Some factors influencing these phenomena could be the interaction between coating components and the temperature and cooling rate. Natural waxes in coating formulations tend to crystallize at low temperatures, and at low crystallization rates, bigger crystals can be formed, acquiring an amorphous and porous structure. All these structural changes could alter the wax stability, becoming more fragile and permeable to water. For this reason, it is possible that the low efficiency of this coating could be attributed to this phenomenon [39–41].

Table 3. Weight loss of cucumbers coated with chitosan, OEO, and carnauba wax and stored at 10 °C for 15 days.

Treatment	Weight Loss at Day 15 (%)
Control	7.11 ± 0.20 ^{b*}
Chitosan	7.18 ± 0.50 ^{b}
Wax	5.48 ± 0.50 ^{a}
Chitosan-wax	7.42 ± 0.21 ^{b}
Chitosan-OEO	7.66 ± 0.29 ^{b}
Wax-OEO	6.93 ± 0.10 ^{a}
Chitosan-OEO-wax	8.03 ± 2.01 ^{b}

Means ± standard deviation, $n = 9$. * Different letters in the same column indicate significant differences ($p \leq 0.05$).

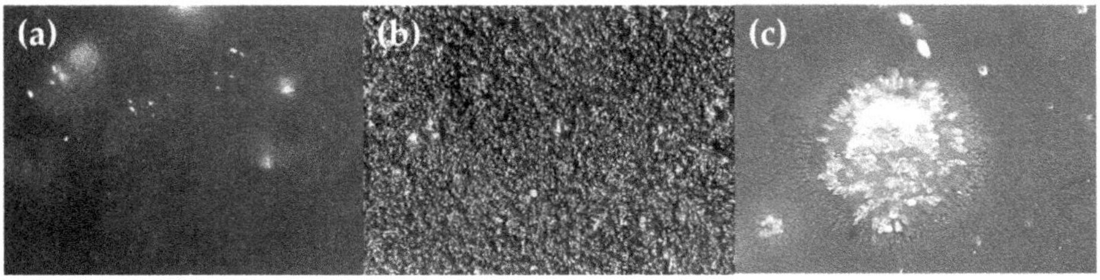

Figure 6. Stereoscopic microphotographs of the wax coated cucumbers stored at (**a**) 25 °C and (**b**,**c**) 10 °C; these images were acquired at a magnitude of 10×.

Won and Min [13] reported that carnauba wax coating significantly reduced weight loss (18.1%) of Satsuma mandarins treated with carnauba wax during storage at 4 and 25 °C for 28 days, compared with control fruit. Also, weight loss was significantly reduced in Valencia oranges after the application of carnauba wax for eight weeks at 4 and 20 °C [14]. Moreover, Miranda et al. [42] observed a similar trend when applying carnauba wax (18%) to papaya fruit, obtaining a 70.6% reduction of weight loss compared to uncoated fruits after 6 days of storage at 20 °C. Therefore, the use of these coatings during storage at higher temperatures can be suggested to avoid crystallization and maintain the response observed in the WVT analysis at 25 °C.

3.2.2. The Microbial Load of Coated Cucumbers

Table 4 shows the bacterial counts of cucumbers stored at 10 °C for 15 days. A decrease of bacterial counts due to the coating application was observed on day 0, and there was a slight increase on day 3, showing all treatments a bacterial growth lower than control. At day 6, wax, wax-OEO, and the bilayer coatings showed an increment of 0.1–1.62 Log CFU·g^{-1} of bacterial load. Moreover, chitosan and chitosan-OEO coatings maintained the cucumber bacterial loads significantly lower than control. However, at day 9 and 12 of storage, bacterial counts of chitosan-OEO increased. During the storage period, the chitosan coating maintained the lowest bacterial growth.

Table 4. Antibacterial activity of chitosan, OEO, and carnauba wax coatings on the mesophilic bacteria load of cucumbers stored at 10 °C for 15 days.

Storage (Days)	Log CFU·g^{-1} *						
	Control	**Chitosan**	**Wax**	**Chitosan-Carnauba Wax**	**Chitosan-OEO**	**Wax-OEO**	**Chitosan-OEO-Wax**
0	6.70 ^{a*}	3.81 ^{b}	4.44 ^{b}	4.62 ^{b}	3.76 ^{b}	4.94 ^{b}	4.80 ^{b}
3	6.0 ^{c}	4.80 ^{a}	5.45 ^{b}	4.90 ^{ab}	5.20 ^{ab}	5.36 ^{ab}	4.96 ^{ab}
6	5.90 ^{b}	4.80 ^{a}	6.97 ^{c}	5.0 ^{a}	4.85 ^{a}	6.98 ^{c}	5.09 ^{a}
9	5.23 ^{bc}	4.70 ^{a}	5.54 ^{c}	4.85 ^{ab}	5.51 ^{c}	5.34 ^{c}	5.11 ^{abc}
12	6.19 ^{c}	5.22 ^{a}	5.39 ^{ab}	5.99 ^{c}	5.73 ^{bc}	6.30 ^{c}	5.80 ^{bc}
15	7.37 ^{d}	5.78 ^{a}	6.90 ^{c}	6.68 ^{bc}	6.39 ^{b}	6.89 ^{c}	6.56 ^{b}

The initial cellular load of mesophilic bacteria in cucumbers before coating was 6.7 Logs CFU·g^{-1}. * Different letters amongst rows indicate significant differences amongst treatments per day ($p \leq 0.05$), $n = 4$.

As explained before, the antibacterial effect of chitosan is attributed to their amino groups; however, it is known that in multicomponent systems some interactions may occur among the functional groups of the components and the coated surface, causing a blockage of active sites and, hence, a reduction of antibacterial activity [28]. It is important to mention that chitosan only affects microorganisms that are in direct contact with its active sites [35]. Based on this, it can be supposed that carnauba wax led to a steric hindrance between the active sites of chitosan and bacteria. On the other hand, OEO components are volatile, and their presence could decrease during storage. Similar to this study, Moreira, Roura and Ponce [8] reported a significant bactericidal effect of chitosan coatings (with reductions of 1.5–2.5 Log CFU·g^{-1}) on aerobic mesophilic bacteria of treated broccoli, compared to uncoated samples. Chiabrando and Giacalone [43] reported that chitosan films caused higher reductions of yeasts (<2.71 Log CFU·g^{-1}) and molds (<1.05 Log CFU·g^{-1}) in fresh-cut nectarines stored at 4 °C, compared to chitosan added with alginate. Similar to our study, Tokatlı and Demirdöven [44] demonstrated that chitosan caused a reduction of 2.71 Log CFU·g^{-1} in aerobic mesophilic bacteria of sweet cherries. Also, Alvarez et al. [45] reported that chitosan coatings (10 and 20 mg·mL^{-1}) reduced aerobic mesophilic counts (2.5–3 Log CFU·g^{-1}) of broccoli florets stored at 7 °C, compared to uncoated florets.

Table 5 shows the mold and yeast counts of coated cucumbers stored at 10 °C for 15 days. It was observed that all treatments showed similar initial counts ($p \geq 0.05$), being different from the higher values of uncoated fruit. However, chitosan-OEO-coated cucumbers showed the lowest counts (the half of the uncoated cucumbers counts). At day 6, no counts of molds and yeasts were detected on chitosan-OEO and chitosan-wax coated cucumbers, while uncoated fruits continue to increase to the end of the storage time. At the end of storage, the most effective treatment against molds and yeasts was chitosan-wax films, followed by chitosan-OEO, wax-OEO, wax, and chitosan-OEO-wax ($p \leq 0.05$).

Table 5. Molds and yeast counts of fresh cucumber coated with chitosan, OEO, and carnauba wax and stored 15 days at 10 °C.

Storage (Days)	Log CFU·g^{-1} *						
	Control	Chitosan	Wax	Chitosan-Wax	Chitosan-OEO	Wax-OEO	Chitosan–OEO-Wax
0	3.58 ^{c*}	2.14 ^{ab}	2.30 ^{ab}	2.18 ^{ab}	1.82 ^{ab}	2.72 ^{bc}	1.87 ^{a}
3	5.07 ^{c}	3.72 ^{a}	5.22 ^{d}	4.12 ^{ab}	4.10 ^{ab}	4.69 ^{bcd}	4.27 ^{abc}
6	4.30 ^{b}	3.0 ^{a}	3.53 ^{a}	–	–	3.37 ^{a}	3 ^{a}
9	4.25 ^{b}	3.20 ^{a}	–	–	4.13 ^{b}	3.07 ^{a}	–
12	4.98 ^{c}	4.12 ^{b}	3 ^{a}	3.0 ^{a}	3 ^{a}	4.05 ^{b}	–
15	5.02 ^{c}	3.94 ^{b}	3.55 ^{ab}	–	3 ^{a}	3.30 ^{ab}	3.69 ^{ab}

The initial load of molds and yeasts in cucumbers before coating was 3.58 Log CFU·g^{-1}. - below 250 CFU·g^{-1}.
* Different letters in rows indicate significant differences amongst treatments per day ($p \leq 0.05$), $n = 4$.

The observed effect of wax coating could be attributed to the oxygen barrier limiting the growth of aerobic microorganisms. Except for specific rumen-inhabiting species, fungi are strict aerobes or can grow only as microaerophiles [46]. Velickova et al. [47] reported that chitosan-beeswax coatings reduced the fungal infection of strawberries stored at 20 °C for 7 days. On the other hand, chitosan coatings (1%) reduced the disease incidence of *B. cinerea* in grape berries by 16.9% and 28.4% after 12 and 24 days post-infection at 10 °C, respectively [48]. In addition, it has been reported that a reduction of fungi infections in plants treated with chitosan, and this effect was attributed to the antifungal properties of chitosan and its ability to stimulate defense mechanisms, such as chitinase and phytoalexins [49]. It is important to mention that no major changes on the firmness of the coated cucumbers were observed during the experiment (Figure S1). Whereas the coated cucumbers showed higher °hue values than the uncoated fruit (Figure S2), no significant changes on L^* were observed among the treatments on control. Finally, the chitosan-wax and wax-OEO showed lower Chroma values than the uncoated fruit and the rest of the treatments.

4. Conclusions

The addition of carnauba wax helped to decrease the WVT of the formulated chitosan films and coatings. Furthermore, the addition of OEO to chitosan films increased the in vitro antimicrobial activity. Also, it can be concluded that the occurrence of physicochemical interactions among the components altered their individual properties. Finally, carnauba wax and carnauba wax-OEO coatings were the most effective in reducing weight loss, while chitosan and chitosan-wax were the most effective to reduce the microbial load of the treated fresh cucumber.

Supplementary Materials: The following are available online at http://www.mdpi.com/2079-6412/10/7/614/s1, Figure S1: Global effect of chitosan, wax, chitosan-wax, chitosan-OEO, wax-OEO, and chitosan-OEO-wax coatings on the firmness of the cucumbers stored at 10 °C for 15 days. Values are means of $n = 3$. Different literals among treatments indicate significant differences ($p \leq 0.05$), Figure S2: Global effect of chitosan, wax, chitosan-wax, chitosan-OEO, wax-OEO, and chitosan-OEO-wax coatings on the surface color of the cucumbers stored at 10 °C for 15 days. Values are means of $n = 3$. Different literals among treatments in the same parameter indicate significant differences ($p \leq 0.05$).

Author Contributions: Conceptualization, M.M.G.-P., L.A.O.-R., B.A.S.-E., M.R.C.-V., G.A.G.-A., J.L.-M., R.M., and J.F.A.-Z.; data curation, M.M.G.-P. and L.A.O.-R.; funding acquisition, J.F.A.-Z.; methodology, M.M.G.-P., B.A.S.-E., M.R.C.-V., and J.F.A.-Z.; supervision, G.A.G.-A., J.L.-M., and J.F.A.-Z.; writing—original draft, M.M.G.-P. and J.F.A.-Z.; writing—review and editing, M.M.G.-P., L.A.O.-R., B.A.S.-E., M.R.C.-V., G.A.G.-A., J.L.-M., R.M., and J.F.A.-Z. All authors have read and agreed to the published version of the manuscript.

Funding: This research received no external funding.

Acknowledgments: M.M.G.-P. acknowledge the national council of science and technology for the fellowship that was received. The technical support of Monica A. Villegas-Ochoa is fully appreciated.

Conflicts of Interest: The authors declare no conflict of interest.

References

1. USDA. Food Data Central. Available online: https://fdc.nal.usda.gov/fdc-app.html#/food-details/168409/nutrients (accessed on 28 April 2020).
2. Barraza-Álvarez, F.V. Calidad morfológica y fisiológica de pepinos cultivados en diferentes concentraciones nutrimentales. *Revista Colombiana de Ciencias Horticolas* **2015**, *9*, 60–71. [CrossRef]
3. Olawuyi, I.F.; Lee, W. Influence of chitosan coating and packaging materials on the quality characteristics of fresh-cut cucumber. *Korean J Food Preserv* **2019**, *26*, 371–380. [CrossRef]
4. Reddy, P.P. Bacterial Diseases and Their Management. In *In Sustainable Crop Protection under Protected Cultivation*; Springer: Berlin/Heidelberg, Germany, 2016; pp. 153–159.
5. Hassan, B.; Chatha, S.A.S.; Hussain, A.I.; Zia, K.M.; Akhtar, N.J.I.J.o.B.M. Recent advances on polysaccharides, lipids and protein based edible films and coatings: A review. *Int. J. Biol. Macromol.* **2018**, *109*, 1095–1107. [CrossRef]
6. Yousuf, B.; Qadri, O.S.; Srivastava, A.K. Recent developments in shelf-life extension of fresh-cut fruits and vegetables by application of different edible coatings: A review. *LWT* **2018**, *89*, 198–209. [CrossRef]
7. Chiabrando, V.; Giacalone, G.J.E.J.o.F. Agriculture. Effects of edible coatings on quality maintenance of fresh-cut nectarines. *Emir. J. Food Agric.* **2016**, 201–207. [CrossRef]
8. Moreira, M.d.R.; Roura, S.I.; Ponce, A. Effectiveness of chitosan edible coatings to improve microbiological and sensory quality of fresh cut broccoli. *LWT Food Sci. Technol.* **2011**, *44*, 2335–2341. [CrossRef]
9. Pavinatto, A.; De Almeida Mattos, A.V.; Malpass, A.C.G.; Okura, M.H.; Balogh, D.T.; Sanfelice, R.C. Coating with chitosan-based edible films for mechanical/biological protection of strawberries. *Int. J. Biol. Macromol.* **2020**, *151*, 1004–1011. [CrossRef] [PubMed]
10. Athmaselvi, K.; Sumitha, P.; Revathy, B.J.I.A. Development of Aloe vera based edible coating for tomato. *Int. Agrophys.* **2013**, *27*, 369–375. [CrossRef]
11. Poverenov, E.; Zaitsev, Y.; Arnon, H.; Granit, R.; Alkalai-Tuvia, S.; Perzelan, Y.; Weinberg, T.; Fallik, E. Effects of a composite chitosan–gelatin edible coating on postharvest quality and storability of red bell peppers. *Postharvest. Biol. Technol.* **2014**, *96*, 106–109. [CrossRef]
12. De Freitas, C.A.S.; de Sousa, P.H.M.; Soares, D.J.; da Silva, J.Y.G.; Benjamin, S.R.; Guedes, M.I.F. Carnauba wax uses in food—A review. *Food Chem.* **2019**, *291*, 38–48. [CrossRef]
13. Won, M.Y.; Min, S.C. Coating Satsuma mandarin using grapefruit seed extract–incorporated carnauba wax for its preservation. *Food Sci. Biotechnol.* **2018**, *27*, 1649–1658. [CrossRef] [PubMed]
14. Motamedi, E.; Nasiri, J.; Malidarreh, T.R.; Kalantari, S.; Naghavi, M.R.; Safari, M. Performance of carnauba wax-nanoclay emulsion coatings on postharvest quality of 'Valencia' orange fruit. *Sci. Hortic* **2018**, *240*, 170–178. [CrossRef]

15. Rodríguez-Rivera, R.; Herrera-González, J.; Mercado-Silva, E.; Vázquez-Barrios, M.; Rivera-Pastrana, D. Natural coatings and essential oils effects on postharvest quality and antioxidant system of organic avocado (Persea americana Mill 'Hass'). In Proceedings of XXX International Horticultural Congress IHC2018: International Symposium on Strategies and Technologies to Maintain Quality 1275, Istanbul, Turkey, 16 August 2018; pp. 147–154.
16. Dhumal, C.V.; Sarkar, P. Composite edible films and coatings from food-grade biopolymers. *J. Food Sci. Technol.* **2018**, *55*, 4369–4383. [CrossRef] [PubMed]
17. Saxena, A.; Sharma, L.; Maity, T. Chapter 34—Enrichment of edible coatings and films with plant extracts or essential oils for the preservation of fruits and vegetables. In *Biopolymer-Based Formulations*; Pal, K., Banerjee, I., Sarkar, P., Kim, D., Deng, W.-P., Dubey, N.K., Majumder, K., Eds.; Elsevier: Amsterdam, The Netherlands, 2020; pp. 859–880. [CrossRef]
18. Rodriguez-Garcia, I.; Cruz-Valenzuela, M.R.; Silva-Espinoza, B.A.; Gonzalez-Aguilar, G.A.; Moctezuma, E.; Gutierrez-Pacheco, M.M.; Tapia-Rodriguez, M.R.; Ortega-Ramirez, L.A.; Ayala-Zavala, J.F. Oregano (*Lippia graveolens*) essential oil added within pectin edible coatings prevents fungal decay and increases the antioxidant capacity of treated tomatoes. *J. Sci. Food Agric.* **2016**, *96*, 3772–3778. [CrossRef] [PubMed]
19. Wang, K.; Jiang, S.; Pu, T.; Fan, L.; Su, F.; Ye, M. Antifungal activity of phenolic monoterpenes and structure-related compounds against plant pathogenic fungi. *Nat. Prod. Res.* **2019**, *33*, 1423–1430. [CrossRef]
20. Kachur, K.; Suntres, Z. The antibacterial properties of phenolic isomers, carvacrol and thymol. *Crit. Rev. Food Sci.* **2019**, 1–12. [CrossRef]
21. Mohammadi, A.; Hashemi, M.; Hosseini, S.M. Chitosan nanoparticles loaded with *Cinnamomum zeylanicum* essential oil enhance the shelf life of cucumber during cold storage. *Postharvest. Biol. Technol.* **2015**, *110*, 203–213. [CrossRef]
22. ORE. ORE-Orégano Orgánico Lippia Graveolens. Available online: http://www.oreganoorganico.com/ (accessed on 15 April 2020).
23. ASTM International. *Standard Test Methods for Water Vapor Transmission of Materials*; ASTM International: West Conshohocken, PA, USA, 2013.
24. Clinical and Laboratory Standards. *Performance Standards for Antimicrobial Susceptibility Testing: Eleventh Informational*; Clinical and Laboratory Standards Institute: Wayne, PA, USA, 2001.
25. BAM-FDA. Bacteriological Analytical Manual. Available online: https://www.fda.gov/food/laboratory-methods-food/bacteriological-analytical-manual-bam#intro (accessed on 10 January 2020).
26. El Knidri, H.; Dahmani, J.; Addaou, A.; Laajeb, A.; Lahsini, A. Rapid and efficient extraction of chitin and chitosan for scale-up production: Effect of process parameters on deacetylation degree and molecular weight. *Int. J. Biol. Macromol.* **2019**, *139*, 1092–1102. [CrossRef] [PubMed]
27. Kaya, M.; Ravikumar, P.; Ilk, S.; Mujtaba, M.; Akyuz, L.; Labidi, J.; Salaberria, A.M.; Cakmak, Y.S.; Erkul, S.K. Production and characterization of chitosan based edible films from Berberis crataegina's fruit extract and seed oil. *Innov. Food Sci. Emerg.* **2018**, *45*, 287–297. [CrossRef]
28. Verlee, A.; Mincke, S.; Stevens, C.V. Recent developments in antibacterial and antifungal chitosan and its derivatives. *Carbohydr. Polym.* **2017**, *164*, 268–283. [CrossRef] [PubMed]
29. Marchese, A.; Arciola, C.R.; Coppo, E.; Barbieri, R.; Barreca, D.; Chebaibi, S.; Sobarzo-Sánchez, E.; Nabavi, S.F.; Nabavi, S.M.; Daglia, M. The natural plant compound carvacrol as an antimicrobial and anti-biofilm agent: Mechanisms, synergies and bio-inspired anti-infective materials. *Biofouling* **2018**, *34*, 630–656. [CrossRef]
30. Ochoa, T.A.; Almendárez, B.E.G.; Reyes, A.A.; Pastrana, D.M.R.; López, G.F.G.; Belloso, O.M.; Regalado-González, C. Design and characterization of corn starch edible films including beeswax and natural antimicrobials. *Food Bioprocess Technol.* **2017**, *10*, 103–114. [CrossRef]
31. Nair, M.S.; Saxena, A.; Kaur, C. Characterization and antifungal activity of pomegranate peel extract and its use in polysaccharide-based edible coatings to extend the shelf-life of capsicum (*Capsicum annuum* L.). *Food Bioprocess Technol.* **2018**, *11*, 1317–1327. [CrossRef]
32. Santos, F.K.G.d.; Silva, K.N.d.O.; Xavier, T.D.N.; Leite, R.H.d.L.; Aroucha, E.M.M. Effect of the addition of carnauba wax on physicochemical properties of chitosan films. *Mater. Res.* **2017**, *20*, 479–484. [CrossRef]
33. Haq, M.A.; Hasnain, A.; Jafri, F.A.; Akbar, M.F.; Khan, A. Characterization of edible gum cordia film: Effects of beeswax. *LWT Food Sci. Technol.* **2016**, *68*, 674–680. [CrossRef]

34. Zhang, R.; Wang, W.; Zhang, H.; Dai, Y.; Dong, H.; Kong, L.; Hou, H.J.C.P. Effects of preparation conditions on the properties of agar/maltodextrin-beeswax pseudo-bilayer films. *Carbohydr. Polym.* **2020**, *236*, 116029. [CrossRef]
35. Elsabee, M.Z.; Morsi, R.E.; Fathy, M. Chapter 44 - Chitosan-Oregano Essential Oil Blends Use as Antimicrobial Packaging Material. In *Antimicrobial Food Packaging*; Barros-Velázquez, J., Ed.; Academic Press: San Diego, CA, USA, 2016; pp. 539–551. [CrossRef]
36. Fernández-Pan, I.; Maté, J.I.; Gardrat, C.; Coma, V. Effect of chitosan molecular weight on the antimicrobial activity and release rate of carvacrol-enriched films. *Food Hydrocoll.* **2015**, *51*, 60–68. [CrossRef]
37. Yuan, G.; Lv, H.; Yang, B.; Chen, X.; Sun, H. Physical properties, antioxidant and antimicrobial activity of chitosan films containing carvacrol and pomegranate peel extract. *Molecules* **2015**, *20*, 11034–11045. [CrossRef]
38. Ruiz-Navajas, Y.; Viuda-Martos, M.; Sendra, E.; Perez-Alvarez, J.; Fernández-López, J. In vitro antibacterial and antioxidant properties of chitosan edible films incorporated with *Thymus moroderi* or *Thymus piperella* essential oils. *Food Control* **2013**, *30*, 386–392. [CrossRef]
39. Bosquez-Molina, E.; Guerrero-Legarreta, I.; Vernon-Carter, E. Moisture barrier properties and morphology of mesquite gum–candelilla wax based edible emulsion coatings. *Food Res. Int.* **2003**, *36*, 885–893. [CrossRef]
40. Endlein, E.; Peleikis, K.H. Natural Waxes—Properties, Compositions and Applications. *SÖFW-Journal* **2011**, *137*.
41. Doan, C.D.; Tavernier, I.; Okuro, P.K.; Dewettinck, K. Internal and external factors affecting the crystallization, gelation and applicability of wax-based oleogels in food industry. *Innov. Sci. Emerg. Technol.* **2018**, *45*, 42–52. [CrossRef]
42. Miranda, M.; Gozalbo, A.M.; Sun, X.; Plotto, A.; Bai, J.; de Assis, O.; Ferreira, M.; Baldwin, E. Effect of mono and bilayers of carnauba wax based nano-emulsion and HPMC coatings on popst-harvest quality of'redtainung'papaya. Proceedings of Embrapa Instrumentação-Artigo em Anais de Congresso (ALICE), São Paulo, Brazil, 3–5 December 2019.
43. Chiabrando, V.; Giacalone, G. Effect of chitosan and sodium alginate edible coatings on the postharvest quality of fresh-cut nectarines during storage. *Int. J. Trop. Subtrop. Hortic.* **2016**, *71*, 79–85. [CrossRef]
44. Tokatlı, K.; Demirdöven, A. Effects of chitosan edible film coatings on the physicochemical and microbiological qualities of sweet cherry (*Prunus avium* L.). *Scientia Hortic* **2020**, *259*, 108656. [CrossRef]
45. Alvarez, M.V.; Ponce, A.G.; Moreira, M.d.R. Antimicrobial efficiency of chitosan coating enriched with bioactive compounds to improve the safety of fresh cut broccoli. *LWT Food Sci. Technol.* **2013**, *50*, 78–87. [CrossRef]
46. Walker, G.M.; White, N.A. Introduction to fungal physiology. In *Fungi: Biology and Applications*; John Wiley & Sons, Inc.: Hoboken, NJ, USA, 2017; pp. 1–35. [CrossRef]
47. Velickova, E.; Winkelhausen, E.; Kuzmanova, S.; Alves, V.D.; Moldão-Martins, M. Impact of chitosan-beeswax edible coatings on the quality of fresh strawberries (*Fragaria ananassa* cv Camarosa) under commercial storage conditions. *LWT Food Sci. Technol.* **2013**, *52*, 80–92. [CrossRef]
48. Kanetis, L.; Exarchou, V.; Charalambous, Z.; Goulas, V. Edible coating composed of chitosan and *Salvia fruticosa* Mill. extract for the control of grey mould of table grapes. *J. Sci. Food Agric.* **2017**, *97*, 452–460. [CrossRef]
49. Kumar, V.; Sangeetha, K.; Ajitha, P.; Aisverya, S.; Sashikala, S.; Sudha, P.J.H.o.B.A.; Applications, M. Chitin and Chitosan: The Defense Booster in Agricultural Field. In *In Handbook of Biopolymers: Advances and Multifaceted Application*; Vijayalakshmi, K.S.K., Ajitha, P., Aisverya, S., Sashikala, S., Sudha, P.N., Eds.; Jenny Stanford Publishing: New York, NY, USA, 2018; p. 93.

coatings

MDPI

Article

Edible Coatings Formulated with Antifungal GRAS Salts to Control Citrus Anthracnose Caused by *Colletotrichum gloeosporioides* and Preserve Postharvest Fruit Quality

Victoria Martínez-Blay, María B. Pérez-Gago, Beatriz de la Fuente, Rosario Carbó and Lluís Palou *

Centre de Tecnologia Postcollita (CTP), Institut Valencià d'Investigacions Agràries (IVIA), Montcada, 46113 València, Spain; martinez_vicbla@externos.gva.es (V.M.-B.); perez_mbe@gva.es (M.B.P.-G.); beattoro@hotmail.com (B.d.l.F.); carbo_con@gva.es (R.C.)

* Correspondence: palou_llu@gva.es; Tel.: +34-963424117

Received: 9 July 2020; Accepted: 23 July 2020; Published: 24 July 2020

Abstract: The in vitro antifungal activity of various generally recognized as safe (GRAS) salts against *Colletotrichum gloeosporioides*, the causal agent of citrus postharvest anthracnose, was evaluated as mycelial growth reduction on potato dextrose agar (PDA) dishes amended with salt aqueous solutions at different concentrations. The most effective treatments [0.2% ammonium carbonate (AC), 2% potassium sorbate (PS), 0.2% potassium carbonate (PC), 0.1% sodium methylparaben (SMP), 0.1% sodium ethylparaben (SEP), 2% sodium benzoate (SB) and 2% potassium silicate (PSi)] were selected as antifungal ingredients of composite edible coatings formulated with hydroxypropyl methylcellulose (HPMC)-beeswax (BW) matrixes. Stable coatings containing these salts were applied in in vivo curative experiments to "Nadorcott" mandarins and "Valencia" oranges artificially inoculated with *C. gloeosporioides* and those containing 2% PS, 2% SB and 2% PSi were the most effective to reduce anthracnose severity with respect to control fruit (up to 70% on mandarins). The effect of these selected coatings on the quality of non-inoculated and cold-stored "Valencia" oranges was determined after 28 and 56 days at 5 °C and 90% RH, followed by 7 days of shelf life at 20 °C. None of the coatings significantly reduced weight loss of coated oranges, but they modified their internal atmosphere, increasing the CO_2 content. Overall, the coatings did not adversely affect the physicochemical and sensory attributes of the fruit.

Keywords: food additives; mandarins; oranges; non-polluting postharvest decay control; cold-stored fruit

1. Introduction

Citrus (*Citrus* spp., Rutaceae) are grown in many countries with tropical and subtropical climate and are among the most important crops produced for human consumption in the world. Total worldwide production of fresh fruits exceeded 130 million tons in 2018 and the most important citrus-producing countries are China, Brazil, India, the United States of America (USA), Spain, Mexico, Egypt, Turkey, Iran, Italy, Argentina, South Africa and Morocco, among others. In terms of international trade, Spain is the leading exporter of citrus fruits for fresh consumption and Valencia is the most important citrus growing region in Spain [1].

Postharvest diseases are one of the most important problems affecting both fresh and juice citrus industries and are mainly caused by fungal pathogens. Fungi can infect the fruit before, during or after harvest, but disease develops when the fruit has been picked, causing important economic losses to the industry in many countries [2–5]. Depending on the climate of the production area where citrus are grown, the importance of the main postharvest diseases varies. In high summer rainfall areas, such as Brazil or Florida, latent infections initiated in the fruit before harvest are the most relevant and are typically caused by the genera *Colletotrichum*, *Lasiodiplodia*, *Phomopsis*, *Alternaria* and *Phytophthora*, among others. In contrast, in areas with low summer rainfall, such as Spain and other Mediterranean countries, California or South Africa, wound pathogens that infect the fruit through injuries inflicted during harvest or after harvest are more prevalent, especially those belonging to the genera *Penicillium*, the cause of green and blue molds, and *Geotrichum*, the cause of sour rot [3,6].

Postharvest anthracnose of citrus fruits, caused by different species of *Colletotrichum*, especially *Colletotrichum gloeosporioides* (Penz.) Penz. & Sacc. (*C. gloeosporioides sensu stricto*), is an important disease in both types of production areas. Citrus anthracnose can also be a field disease, typically caused by *Colletotrichum acutatum* J.H. Simmonds, which can affect leaves and twigs and also cause post bloom fruit drop. *C. gloeosporioides* is a weak pathogen on citrus fruits. Conidia are produced abundantly in acervuli on dead plant parts and are spread over short distances, by rain or overhead irrigation, to the developing fruits. In contrast, ascospores are less numerous but airborne, taking part in long distance dispersal. The spores germinate giving rise to appressoria that generally remain latent on the fruit surface [3,7]. In general, temperatures surrounding 25 °C and relative humidity (RH) higher than 95% are optimal environmental conditions that favor *C. gloeosporioides* germination and appressorium formation [8]. Fruit colonization and decay usually occur after harvest, mostly on tissues weakened due to other factors such as sunburn, overripeness or excessively prolonged cold storage. However, the disease may also develop on early season fruit treated with ethylene for degreening purposes [7]. Symptoms of postharvest anthracnose appear after prolonged wet periods (important in summer-rainfall areas) that favor the production and dispersal of inoculum and the incidence in the field of fruit latent infections. Symptoms associated with weakened fruit are firm, dry, brown to black spots (1.5 mm or more in diameter). Under humid conditions, conidial masses, pink to salmon in color, appear on the lesion surface. Symptoms on ethylene-treated fruits are larger, firm, flat, silver gray lesions with a leathery texture. As the lesion extends, it becomes darker and may affect much of the rind and lead to a brown to black soft rot [3,7].

Postharvest applications of synthetic chemical fungicides have been used for many years as the main tool to control postharvest diseases of citrus fruits, especially green and blue molds. Some of these chemicals, such as thiabendazole (TBZ) and sodium o-phenylphenate (SOPP), have also shown some effect against diseases caused by latent pathogens, particularly against Diplodia and Phomopsis stem-end rots and anthracnose [3,9]. However, the proliferation of resistant fungal strains and the increasing public concerns about the deleterious effect of chemical residues on human health and the environment are factors limiting this practice. Therefore, the adoption of non-polluting alternatives to control citrus postharvest diseases, including anthracnose, is needed [10,11]. Among them, the use of edible coatings formulated with food-grade antifungal compounds allows coating the fruit directly with a thin layer of edible material in order to extend product shelf life [12]. This type of antifungal coatings could be a cost-effective substitute for the use of citrus commercial waxes containing chemical fungicides [13]. Polysaccharides, proteins and lipids are the main ingredients used to formulate composite edible coatings. These ingredients are mixed to reduce gas and water exchange between the fruit and the environment and to improve fruit mechanical and sensorial properties [14].

Antimicrobial ingredients used for the formulation of edible coatings should be classified as generally recognized as safe (GRAS) and approved for their use as food additives by the United States Food and Drug Administration (US FDA) or the European Food Safety Authority (EFSA) [15]. Food additives are widely used as preservatives for controlling food pH, taste or other qualities. Among them, various organic and inorganic salts have antimicrobial action and may offer a good alternative to the use of synthetic fungicides [10,16]. The main advantages of using GRAS salts include their availability, relatively low cost and high solubility in water [17]. In previous works at the IVIA CTP, we have developed and characterized hydroxypropyl methylcellulose (HPMC)-lipid edible coatings containing GRAS salts with activity against fungal pathogens causing postharvest diseases of plums [18,19] or cherry tomatoes [20,21]. Furthermore, on citrus fruits, this type of coatings has also been effective against green and blue molds [10,22–25] and Lasiodiplodia stem-end rot [26]. Antifungal edible coatings have been successfully used to reduce postharvest anthracnose caused by *Colletotrichum* spp. on different fresh fruit commodities, such as mango, avocado, papaya and strawberry [27–29]. However, to our knowledge, there is no information available on the development of edible coatings with antifungal food additives to control citrus postharvest anthracnose caused by *C. gloeosporioides*.

The aims of this research were: (1) to evaluate the in vitro activity of various GRAS salts, at different concentrations, against *C. gloeosporioides* and (2) to develop novel stable HPMC-lipid composite coatings containing the most promising salts and concentrations. The ability of the coatings to control citrus anthracnose was assessed in in vivo experiments with mandarins and oranges artificially inoculated with the pathogen. The effects of selected antifungal edible coatings on physico-chemical and sensorial quality was also determined on oranges stored at 5 °C for up to two months.

2. Materials and Methods

2.1. GRAS Salts

The name, acronym, food additive E-number, molecular formula and molecular weight of the antifungal salts used in this work are given in Table 1. Ammonium carbonate (AC) and ammonium bicarbonate (ABC) were purchased from Thermo Fisher Scientific (Leicestershire, UK); potassium bicarbonate (PBC), potassium carbonate (PC) and sodium benzoate (SB) from Carl Roth® GmbH +Co. KG (Karlsruhe, Germany); potassium sorbate (PS), sodium ethylparaben (SEP) and sodium methylparaben (SMP) from Merck® kGaA (Darmstadt, Germany); potassium silicate (PSi) was acquired from Alfa Aesar® GmbH and Co. KG (Karlsruhe, Germany) and sodium propionate (SP) from Merck Life Science S.L.U (Madrid, Spain).

Table 1. Characteristics of antifungal GRAS salts tested in vitro to inhibit *Colletotrichum gloeosporioides* and in vivo as ingredients of edible coatings to control citrus anthracnose.

GRAS Salt	Acronym	Molecular Formula	E-Number [1]	MW [2]
Ammonium bicarbonate	ABC	NH_4HCO_3	E-503 (ii)	79.06
Ammonium carbonate	AC	$(NH_4)_2CO_3$	E-503 (i)	114.10
Potassium bicarbonate	PBC	$KHCO_3$	E-501 (ii)	100.12
Potassium carbonate	PC	K_2CO_3	E-501 (i)	138.21
Potassium silicate	PSi	K_2SiO_3	E-560	154.26
Potassium sorbate	PS	$C_6H_7O_2K$	E-202	150.22
Sodium benzoate	SB	$C_7H_5O_2Na$	E-211	144.11
Sodium ethylparaben	SEP	$C_9H_9NaO_3$	E-215	188.16
Sodium methylparaben	SMP	$C_8H_7NaO_3$	E-219	174.13
Sodium propionate	SP	CH_3CH_2COONa	E-281	96.06

[1] E-number: codes for substances permitted as food additives within the European Union. [2] Molecular weight (g/mol).

2.2. Fungal Pathogen

The strain *C. gloeosporioides* NAV-1 was used in the present work. It is an isolate obtained from decayed oranges from a local citrus packinghouse in the Valencia region (Spain). This fungal strain was isolated, purified, molecularly identified and maintained in the culture collection of postharvest pathogens of the IVIA CTP. It was also deposited in the Spanish Type Culture Collection (CECT, University of Valencia, Valencia, Spain) with the accession number CECT 21107. Before the experiments, the fungal isolate was incubated on potato dextrose agar (PDA) (Scharlab S.L., Barcelona, Catalonia, Spain) Petri dishes at 25 °C for 7–14 d.

2.3. In Vitro Antifungal Activity of GRAS Salts

The effect of ABC, PBC, PSi, SEP, SMP and SP on radial mycelial growth of *C. gloeosporioides* was evaluated as previously described by Guimarães et al. [26]. In brief, 90-mm plastic Petri dishes with PDA medium were amended, at 40–50 °C, with sterile aqueous solutions of the salts to achieve final concentrations of 0.2%, 1% and 2% (v/v) for ABC, PBC, PSi and SP and of 0.01%, 0.05% and 0.1% (v/v) for the paraben salts. PDA Petri dishes without salt served as controls. The center of each Petri dish was inoculated with a 5-mm diameter mycelial plug, obtained with a sterilized cork-borer, from 7 to 14-d-old cultures of *C. gloeosporioides*. The plates were incubated in a growth chamber at 25 °C in the dark. Radial mycelial growth was determined in each plate by calculating the mean of two perpendicular fungal colony diameters. Results after 3, 5 and 7 d of incubation are presented. Four replicates, each one corresponding with one plate, were used for each salt and concentration. Results are expressed as percentage of mycelial growth inhibition: $[(dc - dt)/dc] \times 100$, where dc = average diameter of the fungal colony on control plates and dt = average diameter of the fungal colony on Petri dishes amended with the salts.

2.4. Preparation of Antifungal Edible Coatings

HPMC-beeswax (BW) composite edible coatings (ECs) were prepared combining the hydrophilic phase (HPMC) with the hydrophobic phase (BW) suspended in water. Glycerol was used as a plasticizer and stearic acid (Panreac Química SA, Barcelona, Catalonia, Spain) as an emulsifier. HPMC (Methocel E15) was purchased from Dow Europe GmbH (Dow Chemical Co., Stade, Germany), glycerol from VWR International (Leuven, Belgium) and BW and stearic acid were supplied by Guinama S.L.U. (La Pobla de Vallbona, Valencia, Spain). All the formulations contained 1.3% HPMC (w/w, wet basis, wb) and 3% BW (wb). Ratios of HPMC-glycerol (2:1) and BW-stearic acid (3:1) and a total solid concentration of 6% were kept constant for all coatings. GRAS salts and concentrations were selected according to the results of the in vitro tests described above and also from the minimum effective concentration reported for *C. gloeosporioides* in previous literature references [30–36]. Then, selected salts and concentrations (w/v) were tested for compatibility with the HPMC-BW coating matrix and only those forming stable emulsions were eventually selected: AC (0.2%), PS (2%), PC (0.2%), SMP (0.1%), SEP (0.1%), SB (2%) and PSi (2%). The pH and viscosity (cP) values of HPMC-BW composite emulsions formulated with these GRAS salts were the following: 6.83 and 46.2 cP, respectively, for AC coating; 6.27 and 51.2 cP for PS coating; 7.15 and 50.0 cP for PC coating; 7.15 and 46.7 cP for SMP coating; 7.03 and 45.9 cP for SEP coating; 6.07 and 46.7 cP for SB coating and 9.50 and 60.0 cP for PSi coating.

Formulations were prepared as previously described by Guimarães et al. [26]. Briefly, an aqueous solution of HPMC (5%, w/w) was prepared by dispersing the HPMC in hot water at 90 °C and later hydration at 20 °C. Water, BW, glycerol and stearic acid were added to the HPMC solution and heated at 98 °C to melt the lipids. In the case of the coating formulated with SP, Tween® 80 (Panreac-Química S.A., Barcelona, Spain) was used as emulsifier instead of stearic acid. Samples were homogenized with a high-shear probe mixer (Ultra-Turrax IKA® model T25, IKA-Werke, Staufen, Germany) for 1 min at 12,000 rpm and 3 min at 22,000 rpm. After adding the corresponding salts, formulations were cooled under agitation (heating magnetic plate, Falc Instruments, F60, Treviglio, Italy) to a temperature lower than 25 °C by placing them in an ice bath and agitation continued for 25 min to ensure complete hydration of the HPMC.

2.5. Fruit

In vivo disease control experiments were conducted with "Nadorcott" hybrid mandarins (*Citrus reticulata* × *Citrus sinensis*) and "Valencia" oranges [*Citrus sinensis* (L.) Osbeck], whereas quality assessments on coated fruits were performed with cold-stored "Valencia" oranges. Mandarins and oranges were collected from commercial orchards in the Valencia area (Spain) and transported to the IVIA CTP facilities. No commercial postharvest treatments were applied. Fruits were selected, randomized, surface disinfected (5-min dips in diluted commercial bleach, 0.5% sodium hypochlorite), rinsed with tap water and allowed to air-dry at room temperature to be used the following day in the experiments.

2.6. *In Vivo Anthracnose Control of Antifungal Coatings*

For inoculation, conidia from 7 to 14-d-old cultures were taken from PDA plates with a sterilized inoculation loop and transferred to a sterile aqueous solution of Tween® 80 (0.05%, w/v). Conidial suspension was filtered through two layers of cheesecloth and the density of the suspension was measured with a hemocytometer. Dilutions with sterile water were done to obtain an exact inoculum density of 2×10^6 spores/mL. Being a weak pathogen on citrus, to prepare the final inoculum of *C. gloeosporioides*, 5 mg/L of cycloheximide (Carl Roth GmbH + Co. KG, Karlsruhe, Germany) were added to the spore suspension in order to inhibit the possible lignification of the inflicted rind wounds.

Each fruit was wounded and inoculated simultaneously, at one point in the equatorial zone, using the tip of a stainless-steel rod (1 mm wide and 2 mm in length) previously immersed in the aforementioned conidial suspension. Inoculated fruits were incubated for 24 h at 25 °C and 95% RH. After this period, fruits were individually coated to assess the curative activity of the coatings. Three hundred microliters of coating material were pipetted onto each fruit and rubbed with gloved hands to simulate the application of coating machinery on roll-conveyors in commercial citrus packinglines [26,37]. Coated fruits were allowed to air-dry at room temperature. Inoculated but uncoated mandarins or oranges served as controls. For each citrus species, four replicates of 10 fruits each were used per treatment. Every trial was repeated once. Treated fruits were arranged on plastic cavity sockets on plastic trays and incubated at 25 °C and 95% RH.

Anthracnose development was assessed as disease severity (lesion diameter) after 7 and 15 d of incubation. Results after 15 d are presented as the percentage of severity reduction with respect to the control treatments.

2.7. *Effect of Coatings on Quality of Cold-Stored Fruit*

HPMC-BW coatings containing the following GRAS salts and concentrations were selected to evaluate their effect on postharvest quality of non-inoculated and cold-stored oranges: PS, PSi and SB, all at 2% (w/v). These coatings were the three most effective among those previously tested for antifungal activity. "Valencia" oranges were selected, washed, coated and stored at 5 °C for four and eight weeks, followed by a shelf-life period of 7 d at 20 °C. Uncoated oranges were used as controls. The following fruit quality attributes were determined at harvest and after cold storage and shelf life.

2.7.1. Weight Loss

Twenty fruits were used to evaluate orange weight loss during storage. After treatment, each fruit was individually numbered and weighed with a calibrated analytical balance (Alessandrini® P30, Modena, Italy). Measurements were performed at the beginning and at the end of each storage period. Results were expressed as the percentage loss of initial weight by using the formula: $\% \text{ WL} = [(\text{Wi} - \text{Wf})/\text{Wi}] \times (100)$, where % WL = percentage of weight loss, Wi = initial fruit weight (g) and Wf = final fruit weight (g).

2.7.2. Fruit Firmness

Firmness of 20 oranges per treatment was determined as percentage of rind deformation, related to initial diameter, with an Instron Universal testing machine (Model 3343, Instron Corp., Canton, MA, USA), according to Valencia-Chamorro et al. [24].

2.7.3. Juice Quality

Soluble solids concentration (SSC, %), titratable acidity (TA, % of citric acid) and maturity index (MI = SSC/TA) were determined as described by Palou et al. [38] in 5 mL juice samples (three replicates of five oranges each per treatment). TA was determined with an automatic titrator (Titrator T50, Mettler Toledo, Switzerland) and SSC was measured using a digital refractometer (model ATC-1, Atago® Co., LTD, Tokyo, Japan).

2.7.4. Internal Gas Concentration

Concentrations of CO_2 and O_2 (%) in the internal cavity of 10 oranges per treatment were determined using a gas chromatograph (GC) (Thermo Trace, Thermo Fisher Scientific, Inc., Waltham, MA, USA) following the methodology described by Valencia-Chamorro et al. [24].

2.7.5. Ethanol Content (EtC) and Acetaldehyde Content (AcC)

The content of these volatile compounds (mg/L) in the headspace of 10-mL vials filled with 5 mL juice samples (three replicates of five oranges each per treatment) was analyzed by gas chromatography according to Valencia-Chamorro et al. [39].

2.7.6. Sensorial Evaluation

Overall taste (1–9 scale, from 1 = very poor to 9 = optimal), the presence of off-flavours (1–5 scale, from 1 = absence to 5 = very pronounced) and external appearance (1–3 scale: 1 = bad, 2 = acceptable and 3 = good) of four coated oranges per treatment were evaluated by a panel of 10 trained tasters following the procedures described by Valencia-Chamorro et al. [24].

2.8. Statistical Analysis

Data from in vitro tests, in vivo trials and fruit quality assessments were subjected to analyses of variance (ANOVA). Since the experiment was not a significant factor, means of repeated experiments are presented. Data on percent inhibition of mycelial growth was subjected to one-way ANOVA with the concentration of the different GRAS salts as dependent variable. Disease reduction with respect to control fruit was calculated as percentage. When appropriate, means separation was performed by Fisher's protected LEAST SIGNIFICANT DIFFERENCE test (LSD, $P = 0.05$). All statistical analyses were performed with the software Statgraphics Centurion XVII (Statgraphics Technologies Inc., The Plains, VA, USA).

3. Results

3.1. In Vitro Antifungal Activity of GRAS Salts

Table 2 shows the radial growth inhibition of colonies of *C. gloeosporioides* compared to control treatment (fungal growth on PDA not amended with GRAS salts) after 3, 5 and 7 d of incubation at 25 °C. Significant differences were found among treatments and the effect of each salt was dependent on the concentration at which it was applied. ABC and SEP were the most effective salts and completely inhibited fungal growth after 7 d of incubation at the intermediate concentrations (1 and 0.05%, respectively). In a second group, SMP also completely inhibited the growth of *C. gloeosporioides* after 7 d at the highest dose of 0.1% and inhibition with PBC exceeded 90% at the highest concentration of 2%. Growth inhibition with 2% SP after 7 d was about 80%, while the least effective GRAS salt was PSi, with 50% of growth inhibition after 7 d at the highest concentration. None of the salts was effective after 7 d of incubation at the lowest concentration tested. SMP, PBC and SP inhibited fungal growth by more than 60% at the intermediate dose tested.

Table 2. Percentage of radial growth inhibition of *Colletotrichum gloeosporioides* on PDA Petri dishes amended with GRAS salts at different concentrations after 3, 5 and 7 d of incubation at 25 °C.

GRAS Salt [1]	Concentration (%)	Inhibition of *C. Gloeosporioides* (%) [2]		
		Day 3	Day 5	Day 7
ABC	0.2	61.44 de	60.49 d	33.43 g
	1	100 a	100 a	100 a
	2	100 a	100 a	100 a
PBC	0.2	20.92 h	26.80 f	23.53 h
	1	98.04 a	87.14 b	76.46 c
	2	100 a	100 a	92.89 b
PSi	0.2	16.99 h	6.43 g	9.22 i
	1	58.50 e	47.32 e	33.72 g
	2	88.56 b	76.57 c	56.96 e
SEP	0.01	40.33 f	36.04 f	35.22 g
	0.05	100 a	100 a	100 a
	0.1	100 a	100 a	100 a
SMP	0.01	30.8 g	29.38 f	29.58 gh
	0.05	100 a	95.72 ab	87.08 b
	0.1	100 a	100 a	100 a
SP	0.2	68.9 d	51.41 de	44.77 f
	1	79.0 c	71.34 c	64.67 d
	2	93.53 ab	76.5 c	79.96 c

[1] See Table 1 for acronym definitions. [2] Colony diameter reduction with respect to control treatments (non-amended PDA dishes). Means in columns with different letters are significantly different by Fisher's protected LSD test ($P < 0.05$) applied after the ANOVA.

3.2. *In Vivo Anthracnose Control of Antifungal Coatings*

The curative effect of coating application [HPMC-BW coatings containing AC (0.2%), PS (2%), PC (0.2%), SMP (0.1%), SEP (0.1%), SB (2%) or PSi (2%)] to control citrus anthracnose after 15 d of incubation at 25 °C and 90% RH is shown in Figure 1. In a first set of experiments with "Nadorcott" mandarins, average data from two trials showed that all inoculated fruits developed decay and all the tested antifungal coatings reduced the severity of the disease (lesion size) between 45% and 70% with respect to uncoated fruits. Coatings formulated with PSi, SB and PS were the most effective, with severity reductions of 70%, 63% and 61%, respectively (Figure 1A). Similarly, in a second set of experiments with "Valencia" oranges, average data from two trials showed a significant reduction in anthracnose severity on coated oranges compared to control fruits. However, this reduction in severity was lower than in "Nadorcott" mandarins, with percentages between 10% and 35%. The most effective coating was that containing SB, followed by those formulated with PS and PSi (Figure 1B). Hence, among all tested coatings, those containing 2% PSi, SB and PS were the most effective to control citrus anthracnose, both in mandarins and oranges, with no significant differences in severity reduction among them.

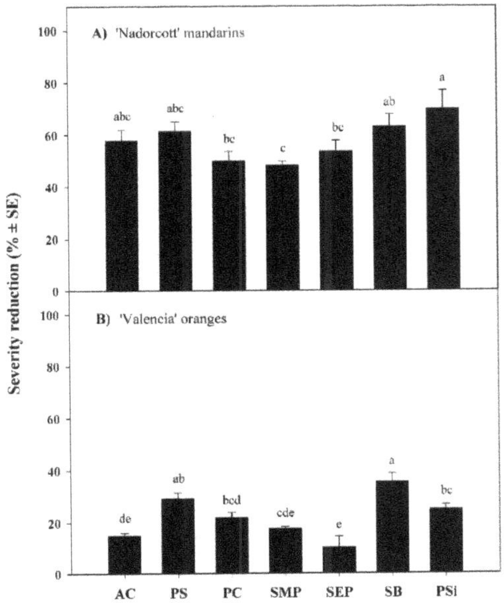

Figure 1. Percentage reduction of anthracnose severity (lesion diameter) with respect to control fruits on "Nadorcott" mandarins (**A**) and "Valencia" oranges (**B**) artificially inoculated with *Colletotrichum gloeosporioides* and coated 24 h later with hydroxypropyl methylcellulose (HPMC)-beeswax (BW) composite edible coatings containing GRAS salts and incubated for 15 d at 25 °C and 90% RH. Represented GRAS salts and concentrations are: 0.2% ammonium carbonate (AC), 2% potassium sorbate (PS), 0.2% potassium carbonate (PC), 0.1% sodium methylparaben (SMP), 0.1% sodium ethylparaben (SEP), 2% sodium benzoate and 2% potassium silicate (PSi). Average data from two trials with each citrus species. In every trial, each treatment was applied to four replications of 10 fruits each. Average severity of uncoated controls was: A) mandarins = 47.29 mm, B) oranges = 15.2 mm. Columns with different letters are significantly different according to Fisher's protected LSD test ($P < 0.05$) applied after the ANOVA.

3.3. *Effect of Coatings on the Quality of Cold-Stored Oranges*

Composite coatings (HPMC-BW) containing PS, PSi or SB at 2% were selected for fruit quality evaluation due to their higher control of anthracnose severity in the previous in vivo trials. Table 3 shows the quality attributes of uncoated (control) and coated "Valencia" oranges at harvest and after cold storage at 5 °C followed by a shelf-life period of 7 d at 20 °C. Weight loss ranged from 2.1% to 2.6% after 28 d of cold storage and from 3% to 4% after 56 d of cold storage, both periods followed by 7 d of shelf life. None of the coatings significantly reduced weight loss compared to uncoated oranges. In the case of the coating formulated with PS, weight loss was even higher than on control fruit after the 28-d storage period. After 56 d, no significant differences were observed between the different coatings and the controls. Fruit firmness, expressed as percentage of rind deformation, decreased after storage (i.e., higher rind deformation) compared to the value at harvest, but no significant differences were found between 28 and 56 d of cold storage and between coated and uncoated fruits, remaining low for all treatments (in the range of 2.5% to 3%).

Table 3. Quality attributes of "Valencia" oranges coated with hydroxypropyl methylcellulose (HPMC)-beeswax (BW) edible composite coatings containing GRAS salts and stored at 5 °C followed by 7 d of shelf life at 20 °C.

Quality Attributes [1]	Storage Conditions and Treatments [2]								
	At Harvest	28 d 5 °C + 7 d 20 °C				56 d 5 °C + 7 d 20 °C			
		Control	HPMC-BW-PS	HPMC-BW-SB	HPMC-BW-PSi	Control	HPMC-BW-PS	HPMC-BW-SB	HPMC-BW-PSi
WL (% ± SE)	–	2.25 ± 0.09 b	2.59 ± 0.10 a	2.24 ± 0.08 b	2.12 ± 0.06 b	3.48 ± 0.12 ab	3.71 ± 0.13 a	3.91 ± 0.22 a	2.97 ± 0.11 b
F (% deformation ± SE)	2.03 ± 0.09	2.58 ± 0.09 a	2.78 ± 0.14 a	2.50 ± 0.07 a	2.56 ± 0.14 a	2.66 ± 0.13 a	2.87 ± 0.15 a	2.47 ± 0.12 a	2.46 ± 0.13 a
SSC (% ± SE)	12.42 ± 0.08	11.70 ± 0.17a	11.37 ± 0.50 ab	10.28 ± 0.55 bc	10.05 ± 0.22 c	11.60 ± 0.26 a	11.20 ± 0.21 a	10.20 ± 0.15 b	10.02 ± 0.23 b
TA (% citric acid ± SE)	1.34 ± 0.01	1.08 ± 0.06 a	1.23 ± 0.10 a	0.97 ± 0.07 a	0.98 ± 0.09 a	0.90 ± 0.04 a	0.84 ± 0.04 a	0.83 ± 0.04 a	0.98 ± 0.04 a
MI (average ± SE)	9.24 ± 0.08	10.85 ± 0.47 a	9.31 ± 0.38 a	10.59 ± 0.24 a	10.36 ± 0.70 a	12.95 ± 0.63 a	13.28 ± 0.07 a	12.28 ± 0.49 a	10.29 ± 0.52 b
EtC (mg/L ± SE)	221.05 ± 10.42	412.35 ± 40.5 c	533.65 ± 16.21 ab	439.51 ± 23.70 bc	628.01 ± 50.27 a	441.49 ± 24.62 c	606.90 ± 29.78 ab	543.17 ± 80.63 bc	699.63 ± 48.35 a
AcC (mg/L ± SE)	3.16 ± 0.21	5.78 ± 0.26 b	6.51 ± 0.22 b	6.17 ± 0.07 b	7.39 ± 0.39 a	5.72 ± 0.23 c	6.88 ± 0.16 b	7.47 ± 0.31 ab	7.72 ± 0.33 a

[1] WL: weight loss, F: firmness, SSC: soluble solids content, TA: titratable acidity, MI: maturity index, EtC: ethanol content, AcC: acetaldehyde content. Means in rows with different letters are significantly different according to Fisher's protected LSD test ($P < 0.05$) applied after the ANOVA. [2] Control: uncoated fruits; HPMC-BW coatings containing: PS: potassium sorbate, SB: sodium benzoate, PSi: potassium silicate.

Regarding juice quality, SSC and TA decreased and MI increased after cold storage and shelf life compared to values at harvest. In general, oranges coated with HPMC-BW-SB and HPMC-BW-PSi coatings had lower SSC than control fruit after both storage periods and no significant differences in TA and MI were observed between control and coated fruits (Table 3). On the other hand, EtC and AcC in "Valencia" oranges increased during storage compared to the values at harvest and reached values after 56 d that ranged from 400 to 700 mg/L of ethanol and from 5 to 8 mg/L of acetaldehyde. Uncoated samples and samples coated with HPMC-BW-PSi had the lower and higher volatile contents, respectively (Table 3).

Figure 2 shows the internal CO_2 and O_2 concentrations of uncoated and coated oranges after storage. At the end of the 28-d and 56-d storage periods, all tested coatings modified the internal atmosphere of "Valencia" oranges with an increase of internal CO_2 and a decrease of internal O_2 compared to uncoated fruit, and the concentrations of internal CO_2 and O_2 on coated oranges reached values around 4–6 and 15–17 kPa, respectively.

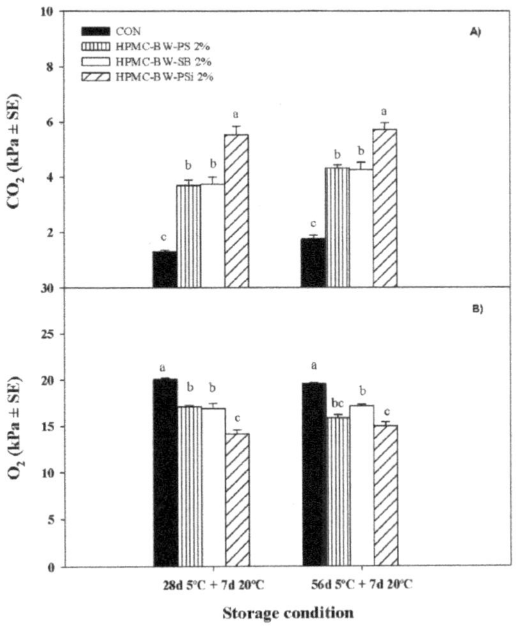

Figure 2. Internal CO_2 (**A**) and O_2 (**B**) concentrations of "Valencia" oranges uncoated (CON) or coated with antifungal hydroxypropyl methylcellulose (HPMC)-beeswax (BW) composite edible coatings and stored at 5 °C followed by 7 d at 20 °C. Coatings contained 2% potassium sorbate (PS), 2% potassium silicate (PSi) or 2% sodium benzoate (SB). For each storage period, columns with different letters are significantly different according to Fisher's protected LSD test ($P < 0.05$) applied after the ANOVA.

The HPMC–BW coatings containing GRAS salts did not modify the flavor of "Valencia" oranges during cold storage, compared to uncoated samples, as determined by the trained judges of the sensory panel (Table 4). Off-flavors ranged between 1.0 (absence) and 1.8 (very slight), with the coating containing 2% SB showing the poorest overall taste and the highest presence of off-flavors after the 56-d storage period, although no significant differences were observed with the control samples ($P > 0.05$). Coating appearance in a 1–3 scale was evaluated according to the presence or absence of cracks, blemishes, stains, and homogeneity of the coating. In general, the appearance of all coated oranges ranged between acceptable and good (1.6–2.7) after both periods of cold storage. However, the incorporation of 2% PS and 2% SB to the HPMC-BW coating matrixes negatively affected the appearance of coated oranges, with the SB-based coating the worst evaluated. The coating containing 2% PSi was the best evaluated in terms of external appearance, without significant differences with the uncoated samples (Table 4).

Table 4. Sensory quality attributes of "Valencia" oranges coated with hydroxypropyl methylcellulose (HPMC)-beeswax (BW) composite edible coatings containing GRAS salts and stored at 5 °C followed by 7 d of shelf life at 20 °C.

Treatments [2]	Storage Conditions and Sensory Attributes [1]					
	28 d 5 °C + 7 d 20 °C			56 d 5 °C + 7 d 20 °C		
	Overall Taste (1–9 Scale) [3]	Off-Flavours (1–5 Scale) [4]	Appearance (1–3 Scale) [5]	Overall Taste (1–9 Scale) [3]	Off-Flavours (1–5 Scale) [4]	Appearance (1–3 Scale) [5]
Control	6.43 ± 0.43 a	1.00 ± 0.00 a	2.43 ± 0.20 a	5.70 ± 0.63 a	1.38 ± 0.16 a	2.38 ± 0.24 a
HPMC-BW-PS	5.71 ± 0.56 a	1.71 ± 0.19 a	1.58 ± 0.20 b	5.50 ± 0.54 a	1.33 ± 0.22 a	1.63 ± 0.16 b
HPMC-BW-SB	6.43 ± 0.20 a	1.00 ± 0.00 a	2.00 ± 0.22 b	5.30 ± 0.58 a	1.80 ± 0.36 a	1.88 ± 0.26 b
HPMC-BW-PSi	5.83 ± 0.28 a	1.57 ± 0.20 a	2.71 ± 0.18 a	5.61 ± 0.41 a	1.33 ± 0.16 a	2.63 ± 0.24 a

[1] Means in columns with different letters are significantly different according to Fisher's protected LSD test applied after the ANOVA ($P < 0.05$). [2] Control: uncoated fruits; HPMC-BW coatings containing: PS: potassium sorbate, SB: sodium benzoate, PSi: potassium silicate. [3] Flavour ranked from 1 (very poor) to 9 (optimum). [4] Off-flavours ranked from 1 (absence) to 5 (presence). [5] Coating/fruit appearance ranked from 1 (bad) to 3 (good).

4. Discussion

The present work highlights the antifungal activity of different GRAS salts or food preservatives against *C. gloeosporioides* and their potential use as ingredients of antifungal composite edible coatings for the control of postharvest anthracnose of citrus fruits. Our in vitro results showed that, among all GRAS salts tested, ABC and SEP were the most effective to inhibit the mycelial growth of *C. gloeosporioides*. Previous works have reported the potential of carbonate salts to reduce the in vitro mycelial development of different *Colletotrichum* spp. Aqueous solutions of the salts AC at 3% [36] and sodium bicarbonate (SBC) at 2% [34] completely inhibited the mycelial growth of *C. gloeosporioides* isolated from papaya, while SBC significantly reduced the mycelial development of the species *Colletotrichum musae* (Berk. & Curtis) Arx. isolated from banana. Similarly, other researchers have also identified ABC as the most effective salt, at all concentrations tested (0.2, 1.0 and 2.0%), to inhibit the growth on PDA dishes of other important postharvest pathogens such as *Monilinia fructicola* (G. Wint.) Honey [19] and *Lasiodiplodia theobromae* (Pat.) Griffon & Maubl. [26]. In addition, various salts, mainly carbonates, were effective to inhibit the in vitro radial growth of *Botrytis cinerea* Pers. [40,41], *Geotrichum citri-aurantii* (Ferraris) Butler [42] and *Penicillium expansum* L. [43]. Likewise, the salt SEP effectively inhibited the growth of different fungi causing major postharvest diseases on fresh horticultural produce, such as *B. cinerea*, *Alternaria alternata* (Fr.) Keiss. [20] and *L. theobromae* [26].

The present results and former research clearly show that some GRAS salts have a broad spectrum of antifungal activity since they are able to inhibit the in vitro growth of a variety of fungal pathogens. The in vitro toxicity of a GRAS salt is influenced by many factors, such as the pathogen species and strain, the salt components (ions) and concentration, the pH, the culture medium and the incubation conditions [15,44,45]. General antifungal mechanisms of action of GRAS salts include the alteration of the integrity and permeability of the fungal cell membranes, interferences in the transport of nutrients and energy metabolism and collapse of hyphae or spores [42,44,46]. It is known that the addition of inorganic or organic salts to the medium modifies its pH and, in general, the antifungal activity of the salt is higher as the pH increases [19,47]. However, the pH alone cannot explain the toxicity of these compounds as different salts with the same pH can affect the same fungal strain differently [26,45]. Moreover, the salt cations and anions also play an important and complex role. In fact, sodium, potassium or ammonium forms of the same salt can show large differences in their toxicity to a particular fungal strain [19,36,48].

Salts and concentrations to be used as ingredients of HPMC-BW edible coatings were selected according to previous in vitro results and their capability to form stable emulsions with appropriate characteristics. For this reason, effective salts, such as ABC, PBC and SP, had to be discarded due to incompatibility with the coating matrix leading to phase separation or undesirable properties of coated fruit. Excessive viscosity, bad surface coverage or appearance of salt residues, blemishes or pitting on the surface of oranges and mandarins were major causes for rejection of some experimental coatings.

Among the salts and concentrations selected to be tested in in vivo trials with "Nadorcott" mandarins and "Valencia" oranges, coatings containing 2% PS, SB and PSi were the most effective in reducing anthracnose severity (up to 70% and 35% on mandarins and oranges, respectively). It is worthy to note that the antifungal effect of salt-containing coatings was higher on the citrus species more susceptible to anthracnose, namely mandarins. As pointed out in the Figure 1 caption, the average anthracnose severity (lesion size) on artificially inoculated and uncoated control fruits was 47.29 mm on mandarins, while it was only 15.2 mm on oranges. Since the inoculum density and the methodology used for artificial inoculation with *C. gloeosporioides* was exactly the same for both types of fruit, these values clearly point out that susceptibility to anthracnose was much lower on oranges than on mandarins. This is a feature that has been previously reported for other citrus postharvest diseases, such as green and blue molds caused by *Penicillium* spp., and can be related with the physical and biochemical properties of the fruit rind [3,6,10]. Previous information on the use of GRAS salts, such as AC, SC, SB or SBC, to control postharvest anthracnose on different fruit crops is available, but in most cases the salts were applied by dipping the fruit in aqueous solutions [30,32–34,36]. On the other hand, although the number of studies is considerably lower, some reports are available on the postharvest use of coatings and waxes to control anthracnose caused by *Colletotrichum* spp. on fresh produce. In general, the most studied coatings are applications of chitosan or other edible matrixes containing essential oils as antifungal ingredients [29,49–51]. Nevertheless, some waxes or coatings formulated with GRAS salts or food additives have also been evaluated for anthracnose reduction on different fresh fruits. For example, AC (3%) and SB (2%) in paraffin wax-based formulations significantly reduced anthracnose in papaya caused by *C. gloeosporioides* [36,52]; the same disease was effectively reduced in papaya by chitosan alone or in combination with 3% AC or 2% SBC during storage at 13.5 °C and 95% RH [35]; the combination of fruit-coating polymers with PS or SB significantly decreased the size of the lesions caused by *C. musae* in wound-inoculated bananas after 7 d of incubation at 25 °C and 90% RH [31]. To our knowledge, this is the first work in which edible coatings formulated with antifungal GRAS salts are applied to control citrus postharvest anthracnose. Within this context, the general antifungal activity of different coating matrices containing food additives reported by these authors working with other fresh commodities is in agreement with the results obtained with citrus fruit in the present study.

HPMC-BW edible coatings formulated with GRAS salts, with similar characteristics to those tested here, have been also evaluated to control other important postharvest diseases of citrus fruits. Among a large variety of HPMC-BW edible films containing GRAS salts, those with PS, SB, SP or their mixtures exhibited a noteworthy in vitro antifungal activity against the citrus pathogens *Penicillium digitatum* (Pers.:Fr.) Sacc. and *Penicillium italicum* Wehmer, and coatings containing these salts were effective in reducing green and blue molds on "Valencia" oranges and "Ortanique" and "Clemenules" mandarins artificially inoculated with these pathogens and incubated at 20 °C for 7 d [22–24]. In another study [26], a large amount of GRAS salts and concentrations were evaluated in in vitro tests against *L. theobromae* and the selected salts were assessed as ingredients of HPMC-BW coatings to control Diplodia stem-end rot caused by this fungus in in vivo experiments. Coatings containing 2% PS, 0.1% SEP, 2% SB and 2% PSi were the most effective, with reductions of disease severity of up to 50% on "Barnfield" oranges and "Ortanique" mandarins artificially inoculated with the pathogen and incubated for 10 d at 28 °C and 90% RH. Moreover, the curative activity of similar composite coatings has also been proved in other fresh fruit pathosystems. Significant reductions of black spot on cherry tomatoes artificially inoculated with *A. alternata* were observed by Fagundes et al. after treatment with HPMC-based coatings containing SB, SEP or SMP [20,21]. In the same way, Karaca et al. [19] reported that HPMC-BW coatings containing PS, SEP, SMP or PSi effectively reduced the incidence and severity of brown rot caused by *M. fructicola* on artificially inoculated plums. Therefore, our present results with citrus anthracnose confirm that HPMC coatings containing PS, SB or PSi are broad-spectrum alternatives for the control of postharvest decay of fresh fruits.

This work and previous research show that the selection of the most appropriate antifungal GRAS salt to confer disease control ability to a coating is greatly dependent on the characteristics of each particular pathosystem, such as the type and properties of the fruit (species and even cultivar) and the particular activity of the salt against the target pathogen. However, other factors are also relevant and often the in vitro antifungal activity cannot anticipate the actual in vivo disease control ability. Tests in Petri dishes allow fully exposure of the fungal structures to the salt, while in in vivo assays, with the salt incorporated into the coating and the coating applied to fruit, the contact between the salt and the pathogen can be limited depending on factors such as the emulsion properties (pH and viscosity), the interaction of the salts with the coating matrix and other components (e.g., emulsifiers and plastisizers), the characteristics of the fruit peel and the environmental storage conditions, among others [19,20,26,39,53,54]. Moreover, in some cases, negative results lead to think that some salts presumably provide additional nutrients or enhanced environmental conditions for the development of the fungal pathogen [20,55]. The mentioned factors may explain why some GRAS salts assayed in this work, such as SMP and SEP, were not as effective in in vivo trials as ingredients of the coatings as they were in the in vitro tests. Therefore, it is very important to adapt the formulations and develop appropriate coatings for each particular fruit species and cultivar and for specific target pathogens and postharvest applications. Postharvest use of coatings containing GRAS salts as ingredients may facilitate a slow diffusion of the active ingredient from the matrix compared to the application of aqueous solutions, which could contribute to extend the antifungal effect on the fruit surface and may also reduce phytotoxicity risks [13,15,56]. Hence, the packingline application in citrus packinghouses of these antifungal edible coatings can be a good alternative for commercial anthracnose control to the application of salt aqueous solutions by drencher, dipping or spraying systems.

After both cold storage periods and shelf life, none of the coatings significantly reduced weight loss with respect to uncoated oranges. Among the different coatings tested, those with SB or PS induced higher weight loss than those with PSi. In general, cellulose-lipid composite coatings are reported to reduce fruit weight loss due to the moisture barrier created by the lipid ingredients (BW, shellac, etc.) of the coating formulation [57]. However, several works have confirmed that the addition of food additives such as GRAS salts to HPMC-based coatings greatly affects the moisture barrier properties of stand-alone films or coatings when applied to different fruits such as cherry tomatoes, citrus or table grapes [22,23,58,59]. Thus, the application of HPMC-BW coatings containing SB or PS did not reduce weight loss of coated "Barnfield" and "Valencia" oranges compared to control samples after cold storage at 5 °C, and in both cases PS was less effective than SB for weight retention [24,26]. However, similar coatings significantly reduced weight loss and maintained firmness of "Clemenules" mandarins without adverse effects on the overall quality of coated fruit [39]. Similar results have been reported in research work with other crops. For example, a HPMC-BW coating containing 2% SB showed potential for postharvest industrial application to cherry tomatoes as it reduced weight loss and controlled black spot during prolonged cold storage [21]. Since the antifungal HPMC-BW coatings developed in this work have not satisfactorily reduced weight loss of cold-stored oranges, probably due to changes originated in the permeability of the cuticle, an aspect to consider for further research might be the modification of their physical characteristics in order to improve water loss control while maintaining their antifungal activity.

In the present work, fruit firmness was not affected by the application of HPMC-BW coatings amended with PS, SB or PSi. Polysaccharides present in the cell wall are responsible for the maintenance of fruit firmness and the degradation of these compounds by hydrolyzing enzymes is the cause of fruit softening during ripening and storage. In addition, the effect of coatings on the maintenance of fruit firmness is usually related to their control of weight loss. According to previous results with HPMC-BW coatings, it seems that the influence of coating on fruit firmness is not only dependent on the coating characteristics but also on the citrus cultivar. For instance, in accordance with our results, Valencia-Chamorro et al. [24] and Guimarães et al. [26] reported that HPMC-BW coatings amended with SB or PS did not affect significantly the firmness of coated "Valencia" and "Barnfield" oranges, respectively. However, "Clemenules" mandarins treated with the same type of coatings were significantly firmer after cold storage and shelf life than uncoated control fruits [39]. This could be related to particular properties of the rind of each citrus species or cultivar by which the effects of coating might be modified. Nevertheless, contradictory results have been reported on the relationship between weight loss and firmness on coated citrus fruits. For instance, while a positive correlation was found by Navarro-Tarazaga et al. [60] for "Ortanique" mandarins, no correlation was observed in studies with "Fortune" mandarins [61], indicating the intervention of multiple factors.

Edible coatings can have the capacity to modify the internal gas composition of fresh fruit in terms of O_2 and CO_2 concentrations [18]. The effect of edible coatings on the delay of changes related to fruit ripening (softening, color change, decrease in acidity, appearance of some physiological disorders, etc.) has been related to the gas barrier exerted on the fruit surface, leading to reductions in the respiration rate and/or weight loss [21,62]. The capacity of an edible coating to create an effective gas barrier depends not only on the coating composition and properties (including the addition of GRAS salts), but also on the fruit, cultivar and storage conditions. In a work conducted by Gunaydin et al. [18], the application of HPMC-BW coatings containing paraben salts resulted in the lowest CO_2 production rates, showing the potential of these coatings as gas barriers on plums. However, Fagundes et al. [21] reported the highest respiration rates in cherry tomatoes coated with HPMC-BW emulsions containing SEP. In the present study, the three selected coatings modified the fruit internal atmosphere, and the internal CO_2 and O_2 levels were significantly higher and lower, respectively, in coated fruit than in control samples, which indicates that the coatings were effective as gas barrier. The CO_2 values (3.5–4.5 kPa) in oranges treated with coatings containing PS or SB were equivalent to those observed in coated "Barnfield" oranges [26] and "Clemenules" mandarins [39], but lower than those observed in "Ortanique" mandarins [25] or "Valencia" oranges [63] coated with similar HPMC-lipid coatings containing GRAS salts. To our knowledge, this is the first report on the potential of HPMC-BW coatings amended with PSi as effective gas barriers for citrus fruits. This coating modified the internal atmosphere of "Valencia" oranges in a greater extend than the rest of tested coatings, to reach internal CO_2 and O_2 values of 5.7 and 15.0 kPa, respectively, which could be due to the interaction of PSi with the coating matrix to form a tider structure with less O_2 permeability.

In general, the creation of a modified atmosphere in coated citrus fruits is accompanied by an increase in the volatiles associated with anaerobic respiration, such as ethanol and acetaldehyde [24,39,61]. This was confirmed in this work, and the coating amended with PSi induced the highest volatile content in accordance with the higher internal CO_2 concentration in the fruit. It is assumable that the specific composition and characteristics of the coatings (i.e., total solid content, viscosity, surface tension, barrier and mechanical properties) may explain the different behavior among coating formulations.

Overall, fruit taste and off-flavors were slightly modified during cold storage. However, there were not significant differences between coated and uncoated "Valencia" oranges after both storage periods and shelf life (Table 4). It is known that citrus off-flavor during storage is due to the accumulation of volatiles, with ethanol the most relevant. Moreover, the application of fruit coatings may enhance this process as they can restrict gas exchange through the peel surface [64,65]. However, in citrus, the level of ethanol in the juice that marks the threshold associated with off-flavor appearance depends on the cultivar and, in general, mandarins are more sensitive to anaerobic conditions and develop off-flavors easier than other citrus fruits [66]. For instance, minimum EtC associated with off-flavors has been reported to be 2000 mg/L in "Valencia" oranges [65], 1000 mg/L in "Clemenules" mandarins [67] and 500–600 mg/L in "Murcott" mandarins [68]. In the present work, EtC levels were much lower (400–700 mg/L) than those reported by other authors, which may explain why the tested coatings did not induce off-flavors.

Regarding appearance, coatings containing 2% SB and 2% PB were the worst evaluated in terms of external aspect (aceptable). In general, HPMC-BW coatings are not characterized for providing significant gloss to coated fruit, generally due to the macro emulsion character of the coating formulation [21,23,24,39]. Moreover, some studies have also reported the presence of white spots on the surface of coated mandarins or oranges that reduced the general good appearance of the fruit when using HPMC-based coatings amended with some GRAS salts, including SB and PS [25,26]. On the other hand, the aspect of oranges treated with coatings amended with PSi was quite good and similar to that of uncoated fruits.

In summary, this research allowed the development of HPMC-BW edible coatings effective to reduce citrus postharvest anthracnose through the addition of antifungal GRAS salts such as PS, SB and PSi to the coating matrix. These coatings significantly reduced anthracnose severity on "Nadorcott" mandarins and "Valencia" oranges artificially inoculated with *C. gloeosporioides* and, although they did not reduce weight loss of coated "Valencia" oranges in comparison with uncoated fruits during cold storage, they modified the internal atmosphere of the fruit without adversely affecting the physicochemical and sensorial attributes of the fruit. Further research should focus on the improvement of physical characteristics of the coatings to enhance water loss control and the external aspect of coated citrus fruit. Information gathered from this study provides a basis for further research into the application of these antifungal coatings and their possible combination with other alternative nonpolluting methods to improve the control of postharvest anthracnose in citrus packinghouses. This is especially important in the case of early-season cultivars of mandarins and oranges that are artificially degreened with exogenous ethylene to obtain the appropriate orange color in the rind before commercialization. Exposure to this gas at typical degreening environmental conditions (20–22 °C and RH > 90%) stimulates the germination of conidia and the formation and germination of appressoria of *C. gloeosporioides* and, thus, exacerbates the development of latent infections and the incidence of citrus postharvest anthracnose [3]. Since citrus degreening are typically performed before fruit handling in the packingline, the application of these antifungal edible coatings in the packingline can be a suitable curative treatment against anthracnose and effectively substitute the use of conventional waxes amended with synthetic chemical fungicides.

Author Contributions: Conceptualization: L.P. and M.B.P.-G.; methodology: L.P. and M.B.P.-G.; validation: V.M.-B., L.P. and M.B.P.-G.; formal analysis: V.M.-B. and B.d.l.F.; investigation: V.M.-B., M.B.P.-G., B.d.l.F., R.C. and L.P.; resources: L.P. and M.B.P.-G.; data curation: V.M.-B., B.d.l.F., R.C. and L.P.; writing—original draft preparation: V.M.-B.; writing—review and editing: V.M.-B., L.P. and M.B.P.-G.; visualization: V.M.-B.; supervision: L.P. and M.B.P.-G.; project administration: L.P. and M.B.P.-G.; funding acquisition: L.P. and M.B.P.-G. All authors have read and agreed to the published version of the manuscript.

Funding: This work was partially funded by the Spanish 'Instituto Nacional de Investigación y Tecnología Agraria y Alimentaria' (INIA; project RTA2012-00061-00-00) and the European Commission (FEDER Program). Dr. Victoria Martínez-Blay research scholarship is supported by the IVIA and the European Social Fund (ESF) (IVIA-ESF 2018 grant no. 24). Fontestad S.A. (Montcada, Valencia) is gratefully acknowledged for providing fruit and technical assistance.

Conflicts of Interest: The authors declare no conflict of interest.

References

1. FAOSTAT, Food and Agriculture Organization of the United Nations. Statistics. Available online: http://www.fao.org/faostat/es/#data/QC (accessed on 17 June 2020).
2. Feng, L.; Wu, F.; Li, J.; Jiang, Y.; Duan, X. Antifungal activities of polyhexamethylene biguanide and polyhexamethylene guanide against the citrus sour rot pathogen *Geotrichum citri-aurantii* in vitro and in vivo. *Postharvest Biol. Technol.* **2011**, *61*, 160–164. [CrossRef]
3. Smilanick, J.L.; Erasmus, A.; Palou, L. Citrus fruits. In *Postharvest Pathology of Fresh Horticultural Produce*; Palou, L., Smilanick, J.L., Eds.; CRC Press, Taylor and Francis Group: Boca Raton, FL, USA, 2020; pp. 3–53.

4. Hao, W.; Zhong, G.; Hu, M.; Luo, J.; Weng, Q.; Rizwan-ul-Haq, M. Control of citrus postharvest green and blue mold and sour rot by tea saponin combined with imazalil and prochloraz. *Postharvest Biol. Technol.* **2010**, *56*, 39–43. [CrossRef]
5. Regnier, T.; Combrinck, S.; Veldman, W.; Du Plooy, W. Application of essential oils as multi-target fungicides for the control of *Geotrichum citri-aurantii* and other postharvest pathogens of citrus. *Ind. Crops Prod.* **2014**, *61*, 151–159. [CrossRef]
6. Palou, L. Penicillium digitatum, *Penicillium italicum* (Green Mold, Blue Mold). In *Postharvest Decay. Control Strategies*; Baños-Bautista, S., Ed.; Elsevier Inc.: London, UK, 2014; pp. 45–102.
7. Timmer, L.W.; Garnsey, S.M.; Graham, J.H. *Compendium of Citrus Diseases*, 2nd ed.; APS Press: St. Paul, MN, USA, 2000; ISBN 9780890542484. (paperback).
8. Siddiqui, Y.; Ali, A. Colletotrichum gloeosporioides (Anthracnose). In *Postharvest Decay*; Bautista-Baños, S., Ed.; Elsevier Inc.: London, UK, 2014; pp. 337–371.
9. Ritenour, M.A.; Zhang, J.; Wardowski, W.F.; Brown, G.E. Postharvest decay control recommendations for Florida citrus fruit. *IFAS Ext.* **2011**, *CIR359A*, 1–6.
10. Palou, L.; Ali, A.; Fallik, E.; Romanazzi, G. GRAS, plant- and animal-derived compounds as alternatives to conventional fungicides for the control of postharvest diseases of fresh horticultural produce. *Postharvest Biol. Technol.* **2016**, *122*, 41–52. [CrossRef]
11. Wisniewski, M.; Droby, S.; Norelli, J.; Liu, J.; Schena, L. Alternative management technologies for postharvest disease control: The journey from simplicity to complexity. *Postharvest Biol. Technol.* **2016**, *122*, 3–10. [CrossRef]
12. Janjarasskul, T.; Krochta, J.M. Edible packaging materials. *Ann. Rev. Food Sci. Technol.* **2010**, *1*, 415–448. [CrossRef]
13. Palou, L.; Valencia-Chamorro, S.A.; Pérez-Gago, M.B. Antifungal edible coatings for fresh citrus fruit: A review. *Coatings* **2015**, *5*, 962–986. [CrossRef]
14. Hernandez-Izquierdo, V.M.; Krochta, J.M. Thermoplastic processing of proteins for film formation—A review. *J. Food Sci.* **2008**, *73*, R30–R39. [CrossRef]
15. Palou, L. Postharvest treatments with GRAS salts to control fresh fruit decay. *Horticulturae* **2018**, *4*, 46. [CrossRef]
16. Arslan, U.; Ilhan, K.; Vardar, C.; Karabulut, O.A. Evaluation of antifungal activity of food additives against soilborne phytopathogenic fungi. *World J. Microbiol. Biotechnol.* **2009**, *25*, 537–543. [CrossRef]
17. Mills, A.A.S.; Platt, H.W.; Hurta, R.A.R. Effect of salt compounds on mycelial growth, sporulation and spore germination of various potato pathogens. *Postharvest Biol. Technol.* **2004**, *34*, 341–350. [CrossRef]
18. Gunaydin, S.; Karaca, H.; Palou, L.; De La Fuente, B.; Pérez-Gago, M.B. Effect of hydroxypropyl methylcellulose-beeswax composite edible coatings formulated with or without antifungal agents on physicochemical properties of plums during cold storage. *J. Food Qual.* **2017**, *2017*, 8573549. [CrossRef]
19. Karaca, H.; Pérez-Gago, M.B.; Taberner, V.; Palou, L. Evaluating food additives as antifungal agents against *Monilinia fructicola* in vitro and in hydroxypropyl methylcellulose-lipid composite edible coatings for plums. *Int. J. Food Microbiol.* **2014**, *179*, 72–79. [CrossRef]
20. Fagundes, C.; Pérez-Gago, M.B.; Monteiro, A.R.; Palou, L. Antifungal activity of food additives in vitro and as ingredients of hydroxypropyl methylcellulose-lipid edible coatings against *Botrytis cinerea* and *Alternaria alternata* on cherry tomato fruit. *Int. J. Food Microbiol.* **2013**, *166*, 391–398. [CrossRef]
21. Fagundes, C.; Palou, L.; Monteiro, A.R.; Pérez-Gago, M.B. Hydroxypropyl methylcellulose-beeswax edible coatings formulated with antifungal food additives to reduce alternaria black spot and maintain postharvest quality of cold-stored cherry tomatoes. *Sci. Hortic.* **2015**, *193*, 249–257. [CrossRef]
22. Valencia-Chamorro, S.A.; Palou, L.; Del Río, M.A.; Pérez-Gago, M.B. Inhibition of *Penicillium digitatum* and *Penicillium italicum* by hydroxypropyl methylcellulose-lipid edible composite films containing food additives with antifungal properties. *J. Agric. Food Chem.* **2008**, *56*, 11270–11278. [CrossRef]
23. Valencia-Chamorro, S.A.; Pérez-Gago, M.B.; del Río, M.A.; Palou, L. Curative and preventive activity of hydroxypropyl methylcellulose-lipid edible composite coatings containing antifungal food additives to control citrus postharvest green and blue molds. *J. Agric. Food Chem.* **2009**, *57*, 2770–2777. [CrossRef]
24. Valencia-Chamorro, S.A.; Pérez-Gago, M.B.; del Río, M.A.; Palou, L. Effect of antifungal hydroxypropyl methylcellulose (HPMC)-lipid edible composite coatings on postharvest decay development and quality attributes of cold-stored "Valencia" oranges. *Postharvest Biol. Technol.* **2009**, *54*, 72–79. [CrossRef]

25. Valencia-Chamorro, S.A.; Pérez-Gago, M.B.; del Río, M.A.; Palou, L. Effect of antifungal hydroxypropyl methylcellulose-lipid edible composite coatings on penicillium decay development and postharvest quality of cold-stored "Ortanique" mandarins. *J. Food Sci.* **2010**, *75*, S418–S426. [CrossRef]
26. Guimarães, J.E.R.; de la Fuente, B.; Pérez-Gago, M.B.; Andradas, C.; Carbó, R.; Mattiuz, B.-H.; Palou, L. Antifungal activity of GRAS salts against *Lasiodiplodia theobromae* in vitro and as ingredients of hydroxypropyl methylcellulose-lipid composite edible coatings to control Diplodia stem-end rot and maintain postharvest quality of citrus fruit. *Int. J. Food Microbiol.* **2019**, *301*, 9–18. [CrossRef] [PubMed]
27. Dong, L.M.; Quyen, N.T.T.; Thuy, D.T.K. Effect of edible coating and antifungal emulsion system on *Colletotrichum acutatum* and shelf life of strawberries. *Vietnam J. Chem.* **2020**, *58*, 237–244. [CrossRef]
28. Iñiguez-Moreno, M.; Ragazzo-Sánchez, J.A.; Barros-Castillo, J.C.; Sandoval-Contreras, T.; Calderón-Santoyo, M. Sodium alginate coatings added with *Meyerozyma caribbica*: Postharvest biocontrol of *Colletotrichum gloeosporioides* in avocado (*Persea americana* Mill. cv. Hass). *Postharvest Biol. Technol.* **2020**, *163*, 111123. [CrossRef]
29. Lima Oliveira, P.D.; de Oliveira, K.Á.R.; dos Santos Vieira, W.A.; Câmara, M.P.S.; de Souza, E.L. Control of anthracnose caused by *Colletotrichum* species in guava, mango and papaya using synergistic combinations of chitosan and *Cymbopogon citratus* (D.C. ex Nees) Stapf. essential oil. *Int. J. Food Microbiol.* **2018**, *266*, 87–94. [CrossRef] [PubMed]
30. Ajith, P.; Lakshmidevi, N. Effects of sodium and potassium salts on *Colletotrichum capsici* incitant of anthracnose on bell pepper. *J. Agric.* **2011**, *7*, 423–430.
31. Al Zaemey, A.B.; Magan, N.; Thompson, A.K. Studies on the effect of fruit-coating polymers and organic acids on growth of *Colletotrichum musae in vitro* and on post-harvest control of anthracnose of bananas. *Mycol. Res.* **1993**, *97*, 1463–1468. [CrossRef]
32. De Costa, D.M.; Gunawardhana, H.M.D.M. Effects of sodium bicarbonate on pathogenicity of *Colletotrichum musae* and potential for controlling postharvest diseases of banana. *Postharvest Biol. Technol.* **2012**, *68*, 54–63. [CrossRef]
33. El-Sayed, M.E.; Layla, F.; Mohamed, O.A.M.; Talaat, I.E.S.; Lobna, R.A.A. Anthracnose disease (*Colletotrichum* sp.) affecting olive fruit quality and its control in Egypt. *Int. J. Agric. Technol.* **2014**, *10*, 1289–1306.
34. Hasan, M.F.; Mahmud, T.M.M.; Kadir, J.; Ding, P.; Zaidul, I.S.M. Sensitivity of *Colletotrichum gloeosporioides* to sodium bicarbonate on the development of anthracnose in papaya (*Carica papaya* L. cv. Frangi). *Aust. J. Crop Sci.* **2012**, *6*, 17–22.
35. Sivakumar, D.; Sultanbawa, Y.; Ranasingh, N.; Kumara, P.; Wijesundera, R.L.C. Effect of the combined application of chitosan and carbonate salts on the incidence of anthracnose and on the quality of papaya during storage. *J. Hortic. Sci. Biotechnol.* **2005**, *80*, 447–452. [CrossRef]
36. Sivakumar, D.; Hewarathgamagae, N.K.; Wijeratnam, R.S.W.; Wijesundera, R.L.C. Effect of ammonium carbonate and sodium bicarbonate on anthracnose of papaya. *Phytoparasitica* **2002**, *30*, 486–492. [CrossRef]
37. Bai, J.; Baldwin, E.A.; Hagenmaier, R.H. Alternatives to shellac coatings provide comparable gloss, internal gas modification, and quality for "Delicious" apple fruit. *HortScience* **2002**, *37*, 559–563. [CrossRef]
38. Palou, L.; Marcilla, A.; Rojas-Argudo, C.; Alonso, M.; Jacas, J.A.; del Río, M.A. Effects of X-ray irradiation and sodium carbonate treatments on postharvest *Penicillium* decay and quality attributes of clementine mandarins. *Postharvest Biol. Technol.* **2007**, *46*, 252–261. [CrossRef]
39. Valencia-Chamorro, S.A.; Palou, L.; del Río, M.A.; Pérez-Gago, M.B. Performance of hydroxypropyl methylcellulose (HPMC)-lipid edible coatings with antifungal food additives during cold storage of "Clemenules" mandarins. *LWT Food Sci. Technol.* **2011**, *44*, 2342–2348. [CrossRef]
40. Alaoui, F.T.; Askarne, L.; Boubaker, H.; Boudyach, E.H.; Ait Ben Aoumar, A. Control of gray mold disease of tomato by postharvest application of organic acids and salts. *Plant Pathol. J.* **2017**, *16*, 62–72. [CrossRef]
41. Youssef, K.; Roberto, S.R. Salt strategies to control *Botrytis* mold of "Benitaka" table grapes and to maintain fruit quality during storage. *Postharvest Biol. Technol.* **2014**, *95*, 95–102. [CrossRef]
42. Talibi, I.; Askarne, L.; Boubaker, H.; Boudyach, E.H.; Aoumar, A.A.B. *In vitro* and *in vivo* antifungal activities of organic and inorganic salts against citrus sour rot agent *Geotrichum candidum*. *Plant Pathol. J.* **2011**, *10*, 138–145. [CrossRef]
43. Lai, T.; Bai, X.; Wang, Y.; Zhou, J.; Shi, N.; Zhou, T. Inhibitory effect of exogenous sodium bicarbonate on development and pathogenicity of postharvest disease *Penicillium expansum*. *Sci. Hortic.* **2015**, *187*, 108–114. [CrossRef]

44. Ricke, S. Perspectives on the use of organic acids and short chain fatty acids as antimicrobials. *Poult. Sci.* **2003**, *82*, 632–639. [CrossRef]
45. Hervieux, V.; Yaganza, E.S.; Arul, J.; Tweddell, R.J. Effect of organic and inorganic salts on the development of *Helminthosporium solani*, the causal agent of potato silver scurf. *Plant Dis.* **2002**, *86*, 1014–1018. [CrossRef]
46. Lucera, A.; Costa, C.; Conte, A.; Del Nobile, M.A. Food applications of natural antimicrobial compounds. *Front. Microbiol.* **2012**, *3*, 287. [CrossRef]
47. Xu, W.Q.; Hang, Y.D. Inhibition of *Geotrichum candidum* by bicarbonate. *J. Appl. Microbiol. Biotechnol.* **1989**, *5*, 109–113. [CrossRef]
48. DePasquale, D.A.; Montville, T.J. Mechanism by which ammonium bicarbonate and ammonium sulfate inhibit mycotoxigenic fungi. *Appl. Environ. Microbiol.* **1990**, *56*, 3711–3717. [CrossRef] [PubMed]
49. Maqbool, M.; Ali, A.; Alderson, P.G.; Mohamed, M.T.M.; Siddiqui, Y.; Zahid, N. Postharvest application of gum arabic and essential oils for controlling anthracnose and quality of banana and papaya during cold storage. *Postharvest Biol. Technol.* **2011**, *62*, 71–76. [CrossRef]
50. Bill, M.; Sivakumar, D.; Korsten, L.; Thompson, A.K. The efficacy of combined application of edible coatings and thyme oil in inducing resistance components in avocado (*Persea americana* Mill.) against anthracnose during post-harvest storage. *Crop Prot.* **2014**, *64*, 159–167. [CrossRef]
51. Ali, A.; Hei, G.K.; Keat, Y.W. Efficacy of ginger oil and extract combined with gum arabic on anthracnose and quality of papaya fruit during cold storage. *J. Food Sci. Technol.* **2016**, *53*, 1435–1444. [CrossRef]
52. Gamagae, S.U.; Sivakumar, D.; Wijesundera, R.L.C. Evaluation of post-harvest application of sodium bicarbonate-incorporated wax formulation and *Candida oleophila* for the control of anthracnose of papaya. *Crop Prot.* **2004**, *23*, 575–579. [CrossRef]
53. Valencia-Chamorro, S.A.; Palou, L.; del Río, M.A.; Pérez-Gago, M.B. Antimicrobial edible films and coatings for fresh and minimally processed fruits and vegetables: A review. *Crit. Rev. Food Sci. Nutr.* **2011**, *51*, 872–900. [CrossRef]
54. Valdés, A.; Ramos, M.; Beltrán, A.; Jiménez, A.; Garrigós, M.C. State of the art of antimicrobial edible coatings for food packaging applications. *Coatings* **2017**, *7*, 56. [CrossRef]
55. Palou, L.; Usall, J.; Smilanick, J.L.; Aguilar, M.J.; Viñas, I. Evaluation of food additives and low-toxicity compounds as alternative chemicals for the control of *Penicillium digitatum* and *Penicillium italicum* on citrus fruit. *Pest Manag. Sci.* **2002**, *58*, 459–466. [CrossRef]
56. Vargas, M.; Pastor, C.; Chiralt, A.; McClements, D.J.; González-Martínez, C. Recent advances in edible coatings for fresh and minimally processed fruits. *Crit. Rev. Food Sci. Nutr.* **2008**, *48*, 496–511. [CrossRef]
57. Contreras-Oliva, A.; Rojas-Argudo, C.; Pérez-Gago, M.B. Effect of solid content and composition of hydroxypropyl methylcellulose-lipid edible coatings on physicochemical, sensory and nutritional quality of "Valencia" oranges. *Int. J. Food Sci. Technol.* **2011**, *46*, 2437–2445. [CrossRef]
58. Fagundes, C.; Palou, L.; Monteiro, A.R.; Pérez-Gago, M.B. Effect of antifungal hydroxypropyl methylcellulose-beeswax edible coatings on gray mold development and quality attributes of cold-stored cherry tomato fruit. *Postharvest Biol. Technol.* **2014**, *92*, 1–8. [CrossRef]
59. Pastor, C.; Sánchez-González, L.; Marcilla, A.; Chiralt, A.; Cháfer, M.; González-Martínez, C. Quality and safety of table grapes coated with hydroxypropylmethylcellulose edible coatings containing propolis extract. *Postharvest Biol. Technol.* **2011**, *60*, 64–70. [CrossRef]
60. Navarro-Tarazaga, M.L.; del Río, M.A.; Krochta, J.M.; Pérez-Gago, M.B. Fatty acid effect on hydroxypropyl methylcellulose-beeswax edible film properties and postharvest quality of coated "Ortanique" mandarins. *J. Agric. Food Chem.* **2008**, *56*, 10689–10696. [CrossRef]
61. Pérez-Gago, M.B.; Rojas, C.; del Río, M.A. Effect of lipid type and amount of edible hydroxypropyl methylcellulose-lipid composite coatings used to protect postharvest quality of mandarins cv. Fortune. *J. Food Sci.* **2002**, *67*, 2903–2910. [CrossRef]
62. Valero, D.; Díaz-Mula, H.M.; Zapata, P.J.; Guillén, F.; Martínez-Romero, D.; Castillo, S.; Serrano, M. Effects of alginate edible coating on preserving fruit quality in four plum cultivars during postharvest storage. *Postharvest Biol. Technol.* **2013**, *77*, 1–6. [CrossRef]
63. Navarro-Tarazaga, M.L.; Pérez-Gago, M.B.; Goodner, K.L.; Plotto, A. A new composite coating containing HPMC, beeswax, and shellac for "Valencia" oranges and "Marisol" tangerines. *Proc. Fla. State Hortic. Soc.* **2007**, *120*, 228–234.

64. Tietel, Z.; Plotto, A.; Fallik, E.; Lewinsohn, E.; Porat, R. Taste and aroma of fresh and stored mandarins. *J. Sci. Food Agric.* **2011**, *91*, 14–23. [CrossRef]
65. Ke, D.; Kader, A.A. Tolerance of "Valencia" oranges to controlled atmospheres, as determined by physiological responses and quality attributes. *J. Am. Soc. Hortic. Sci.* **1990**, *115*, 779–783. [CrossRef]
66. Shi, J.X.; Goldschmidt, E.E.; Goren, R.; Porat, R. Molecular, biochemical and anatomical factors governing ethanol fermentation metabolism and accumulation of off-flavors in mandarins and grapefruit. *Postharvest Biol. Technol.* **2007**, *46*, 242–251. [CrossRef]
67. Navarro-Tarazaga, M.L.; Pérez-Gago, M.B. Effect of edible coatings on quality of mandarins cv. Clemenules. *Proc. Fla. State Hortic. Soc.* **2006**, *119*, 350–352.
68. Shi, J.X.; Porat, R.; Goren, R.; Goldschmidt, E.E. Physiological responses of "Murcott" mandarins and 'Star Ruby'grapefruit to anaerobic stress conditions and their relation to fruit taste, quality and emission of off-flavor volatiles. *Postharvest Biol. Technol.* **2005**, *38*, 99–105. [CrossRef]

coatings

MDPI

Article

Antifungal Polyvinyl Alcohol Coatings Incorporating Carvacrol for the Postharvest Preservation of *Golden Delicious* Apple

Mayra Sapper, Maria Eugenia Martin-Esparza, Amparo Chiralt and Chelo Gonzalez Martinez *

Instituto Universitario de Ingeniería de Alimentos para el Desarrollo, Ciudad Politécnica de la Innovación, Universitat Politécnica de Valencia, Camino de Vera, s/n, 46022 Valencia, Spain; maysap@etsiamn.upv.es (M.S.); eesparza@tal.upv.es (M.E.M.-E.); dchiralt@tal.upv.es (A.C.)

* Correspondence: cgonza@tal.upv.es

Received: 7 October 2020; Accepted: 21 October 2020; Published: 26 October 2020

Abstract: Different polyvinyl alcohol (PVA) coating formulations incorporating starch (S) and carvacrol (C) as the active agent were applied to *Golden Delicious* apples to evaluate their effectiveness at controlling weight loss, respiration rate, fruit firmness, and fungal decay against *B. cinerea* and *P. expansum* throughout storage time. Moreover, the impact of these coatings on the sensory attributes of the fruit was also analyzed. The application of the coatings did not notably affect the weight loss, firmness changes, or respiration pathway of apples, probably due to the low solid surface density of the coatings. Nevertheless, they exhibited a highly efficient disease control against both black and green mold growths, as a function of the carvacrol content and distribution in the films. The sensory analysis revealed the great persistence of the carvacrol aroma and flavor in the coated samples, which negatively impact the acceptability of the coated products.

Keywords: PVA; starch; weight loss; firmness; respiration rate; *P. expansum*; *B. cinerea*

1. Introduction

Postharvest blue mold, caused by *Penicillium expansum*, and gray mold, caused by *Botrytis cinerea*, are two of the most common fungal diseases in apples, pears, and a number of other pectin-rich fruits [1–3]. Initial infection most often occurs at sites of fruit injury, such as bruises, natural openings, or puncture wounds. Although infections may start in the field, infected spots often become evident post-harvest, and expand while fruits are in storage due to the combination of intrinsic factors, such as the high sugar content, water activity, and ideal pH; together with favorable environmental conditions, such as low temperatures and high humidity, which permit the postharvest deterioration of the harvested fruit, this causes considerable economic losses [2,4,5]. In pome fruits, disease symptoms include soft, light brown watery lesions. Sometimes, infection can develop from placing rotted fruit next to healthy fruit, spoiling entire lots [6].

Traditionally, the postharvest management of fresh fruit and vegetable decay involves the use of synthetic chemical fungicides. However, growing public concern over the health and environmental hazards associated with the increased levels of chemical fungicides and the lack of approval for the renewal of some of the most effective active molecules has led to the development of safe, alternative, and natural methods of post-harvest disease control. The application of active coatings using antimicrobial compounds of natural origin for fruit preservation purposes could solve some of the challenges associated with stable quality, nutritional value, health safety, and economic production costs [7]. These coatings can modify the internal gas composition and reduce the water loss through the regulation of O_2, CO_2, and water vapor exchange between the fruit and the external atmosphere.

These modifications could affect the physiological behavior of the coated products, associated with the shelf life of produce, and, in some cases, even modify their characteristics prior to consumption [8].

The incorporation of active ingredients, such as antimicrobial agents (essential oils or their components) into the coating matrix represents an additional advantage, since it permits the reduction of the doses of the active compounds while maintaining their effectiveness [9]. Moreover, the application of this technology could minimize one of the major drawbacks of using essential oil-based compounds, such as its potential phytotoxicity; this shows up as spotting on fruit skin, leading to a loss of marketability and to its strong aroma/flavor, which could affect the organoleptic properties of the product, leading to sensory incompatibilities of the selected active compound with the target fruit.

Of the natural antimicrobial compounds, thymol, eugenol, carvacrol and other terpenoids, and phenolic acids from plant essential oils have been widely reported to effectively inhibit mycelial growth and spore germination through fungistatic and/or fungicidal actions against several microorganisms in both in vitro and in vivo studies [10–13]. Of them, the monoterpenoid phenol carvacrol (one of the major constituents of oregano and thyme essential oils) is considered as a safe food additive in Europe and the USA due to the "generally recognized as safe" status [14]. Carvacrol exhibited antibacterial and antioxidant activity and several studies have demonstrated its effectivity against several food-related spoilage fungi such as *Fusarium* spp., several *Aspergillus* and *Penicillium* strains, *Cladosporium* spp., *Botrytis cinerea*, *Fusarium oxysporum* and *Rhizopus oryzae*, etc. [15–20].

Starch, widely used as a food coating/packaging polymer, is available from diverse plant sources, low cost, and biodegradable. Due to their hydrophilic nature, starch films are highly water sensitive and exhibit poor water vapor barrier properties [21]. To overcome these problems, starch is often blended with other biopolymers in order to obtain coatings/films with enhanced properties. To this end, different studies into starch-polyvinyl alcohol (PVA) blend films have been carried out [22–25]. PVA is a hydrophilic, nontoxic synthetic polymer, biocompatible and biodegradable, resulting in an eco-friendly material [26]. PVA has also received FDA approval for close contact with food products, making it widely used as a cold and hot water-soluble film for diverse packaging applications, including food products, detergents, pharmaceuticals, and agricultural chemicals. Recently, PVA has been submitted to FDA approval for use as a component of a water-soluble edible film containing dry food ingredients (GRAS Notice no. 676, 2018). PVA films have good oxygen and aroma barrier properties, good transparency and high tensile strength, and flexibility. Some authors reported that blend films based on starch-PVA presented several advantages over pure starch films, due to the formation of interpenetrated polymer networks with positive effects on the mechanical and water barrier properties of the composite films [22].

In this study, PVA-starch coatings incorporating carvacrol were applied to apples to evaluate: (1) The postharvest behavior of coated fruit in terms of weight loss, respiration rates, and fruit firmness, (2) the antifungal efficacy of these coatings applied as a curative treatment against *B. cinerea* and *P. expansum*, (3) the sensory acceptance of the coated product.

2. Materials and Methods

2.1. Materials

Polyvinyl alcohol (PVA) (M_w: 13,000–23,000, 87%–89% hydrolyzed) was purchased from Sigma-Aldrich Química S.L. (Madrid, Spain), native potato starch was supplied by Roquette Laisa España S.A. (Benifaió, Valencia, Spain), and carvacrol (C) was provided by Sigma-Aldrich (Steinheim, Germany). Glycerol, used as a starch plasticizer and methanol were supplied by Panreac Química S.A. (Castellar de Vallès, Barcelona, Spain).

Apples (*Malus domestica Borkh cv. Golden Delicious*) were purchased from a local packinghouse (Valencia, Spain). The fruit was chosen according to its uniform shape, size, color, and the absence of surface defects.

2.2. Preparation of the Coatings Forming Dispersions and Films

The coating forming dispersions (CFD) were prepared on the basis of previous studies [22]. Thus, starch (2.5% *w/w*) was dispersed in distilled water and kept at 95 °C for 30 min to induce complete starch gelatinization. Meanwhile, 2.5% (*w/w*) PVA aqueous dispersion was obtained under stirring at 90 °C for 30 min. Both solutions were cooled down to reach room temperature and afterwards, glycerol was added to the starch dispersion (0.25 *g/g* of polymer). Carvacrol, used as an antifungal agent, was incorporated into the PVA dispersion (40% or 80% with respect to the PVA) and homogenized for 4 min at 12,500 rpm using an Ultra Turrax rotor-stator homogenizer (DI 25 Basic, IKA®-Werke GmbH & Co. KG, Staufen, Germany). The starch and PVA dispersions were mixed in the adequate proportion to obtain the different CFD. Table 1 shows the different CFD formulations and their respective solid composition.

Table 1. Mass fraction (X) (g/100) and viscosity of the different components in each coating forming dispersion (CFD) and carvacrol retention percentage in the dry films. Mean values (and standard deviation). (S: Starch; PVA: Polyvinyl alcohol; Gly: Glycerol; and C: Carvacrol).

Formulation	CFD					Film
	X_{PVA}	X_S	X_{GLY}	X_C	μ (mPa·s)	Carvacrol Retention (%)
100PVA:0S	2.5	-	-	-	2.92 (0.12) ^{a}	-
100PVA:0S-C40	2.5	-	-	1	2.89 (0.09) ^{a}	59 (2) ^{a}
75PVA:25S-C40	1.875	0.625	0.156	0.75	3.51 (0.02) ^{b}	44.3 (0.4) ^{c}
50PVA:50S-C40	1.25	1.25	0.312	0.5	5.33 (0.05) ^{c}	47 (1) ^{c}
50PVA:50S-C80	1.25	1.25	0.312	1	6.04 (0.04) ^{d}	54 (2) ^{b}

$^{a–d}$: Different superscript letters within the same column indicate significant differences among formulations ($p < 0.05$).

Standalone films were also obtained in order to evaluate the final carvacrol content expected in the coatings after their drying. To this end, a mass of the formulations containing 1.5 g of total solids was spread evenly onto Teflon casting plates (150 mm in diameter) to provide a density of solid of 84 g/m^2. The films were dried under natural convection for approximately 48 h at 25 °C and 45% relative humidity (RH). After drying, the films were peeled off the casting surface and conditioned at 0% RH (using P_2O_5) and at 25 °C.

2.3. Characterization of Coating Forming Dispersions and Carvacrol Retention in the Films

2.3.1. Rheological Behavior of the Dispersions

The rheological behavior of the different formulations was analyzed in triplicate at 25 °C by means of a rotational rheometer (HAAKE Rheostress 1, Thermo Electric Corporation, Karlsruhe, Germany) by using a sensor system of coaxial cylinders, type Z34DIN Ti. Measurements were taken between 0–150 s^{-1} where Newtonian behavior could be assumed and the viscosity values (μ) of the dispersions were determined.

2.3.2. Carvacrol Retention in the Films

To quantify the retention of the active compound during film formation, a known mass of dried film was placed in triplicate in amber vials containing 15 mL of an aqueous solution of methanol 50% (*v/v*), hermetically sealed and kept under stirring at 300 rpm for 24 h at 25 °C to promote carvacrol extraction. Subsequently, aliquots of the sample extract were measured as to the absorbance (A) at 275 nm, using a spectrophotometer (Evolution 201 UV-Vis, Thermo Fisher Scientific Inc., Shanghai, China), as previously described by [27]. The carvacrol concentration (C) in the films was determined by means of a calibration curve obtained with the carvacrol solutions in the same solvent containing between 1118 and 71.28 μg/mL ($C = 0.014A + 0.0048$; $r^2 = 0.9995$). As blank samples, the extract of the corresponding film without carvacrol was considered (2.3.2).

The carvacrol distribution in the films was analyzed through the microstructure of the film's cross-sections. To this end, the film samples were previously conditioned in desiccators containing P_2O_5 in order to eliminate the water content; then, they were immersed in liquid nitrogen to obtain cryo-fractured cross-sections. All of the samples were mounted on copper stubs and platinum coated. The images were obtained by Field Emission Scanning Electron Microscopy (FESEM) (ZEISS®, model ULTRA 55, Oberkochen, Germany), using an accelerating voltage of 2 kV.

2.4. Quality of Coated Fruit

The apples were cleaned and disinfected by immersion in a 1% sodium hypochlorite solution, thoroughly rinsed with tap water, and air-dried at room temperature before the coating application. The samples were dipped in the different CFD for 3 s, while the non-coated samples (controls) were immersed in a water bath. The CFD were allowed to drip off and afterwards, the applied coatings were dried by natural convection for about 24 h at room temperature and stored at 25 °C and 65% RH for 14 days. Five pieces of fruit were considered for each formulation.

2.4.1. Surface Density of Solids (SDS)

The SDS of each coating was evaluated by weighing each fruit with a precision balance (Kern PFB 120-3, Balinguen, Germany) before and after the coating application to obtain the CFD adhered mass, as it has been reported by other authors [21]. To calculate the total adhered solids, the mass fraction of each CFD was considered and the *SDS* (g·m^{-2}) was calculated by applying Equation (1), according to [28]:

$$SDS = \frac{(m_C - m_0)\cdot X_{SCFS}}{m_0}\cdot \rho \cdot \frac{1}{S_e} \quad (1)$$

where m_C is the mass of the coated apple, m_0 is the mass of the uncoated apple, X_{SCFS} is the mass fraction of the solids of the CFD (g solids/g solution), and ρ is the apple density (0.9 g·cm^{-3}). To obtain the specific surface ($Se = 6/d$, m^2 particles m^{-3} fruit), the average diameter (d) was calculated considering a spherical geometry for the fruit.

2.4.2. Weight Loss Rate

Fruit weight loss during storage was determined using an analytical balance (ME235P, Sartorius, Wertheim, Germany) before and after 3, 7, and 14 days of the storage period. The mass loss was referred to the initial mass of each fruit, and the results were expressed as a relative mass loss rate (day^{-1}), which was obtained from the slope of the fitted straight line to the relative weight loss versus time. Five repetitions were considered for each formulation (coated and non-coated).

2.4.3. Respiration Rates

A closed system was used to measure the respiration rate, according to the method proposed by [29], with some modifications. Thus, two apples were placed into 0.940 L hermetic glass jars with a septum in the lid for sampling gas in the headspace at different times. Gas sampling was carried out every 30 min for 4 h by means of a needle connected to a gas analyzer (CheckMate 9900 PBI Dansensor, Ringsted, Denmark). Two replicates per treatment were performed after 3, 7, and 14 days of the storage period. The respiration rate (R_i) of the samples in terms of CO_2 generation and O_2 consumption was determined from the slope of the fitted linear equation, according to Equation (2). The respiration quotient (RQ) was determined as the ratio between CO_2 production and the O_2 consumption.

$$y_{it} = y_{i0} \pm 100\cdot R_i\cdot \frac{M}{V}\cdot t \quad (2)$$

where y_{it} is the gas concentration (%O_2, %CO_2) at time t, y_{i0} is the initial gas concentration, R_i is the respiration rate (mL·kg^{-1}·h^{-1}), M is the mass of the samples, V is the volume (mL) of headspace, and t is time.

2.4.4. Fruit Firmness

Fruit firmness was measured through a puncture test using a texture analyzer (Stable Micro Systems, TA.XT plus, Haslemere, England) with a 50 N load cell equipped with an 11-mm diameter cylindrical probe, applying a modification of the method proposed by [30]. A small skin area was removed from four opposite sides of each fruit in the equatorial zone where the puncture test was carried out. The probe penetrated the flesh at 10 mm·min^{-1} and the force and distance at the break point of the flesh (Fmax, N, dmax, mm) were determined. Four measurements were taken around the equatorial plane of the apple in five different samples for each treatment at 0 and after 14 days of storage.

2.5. *In Vivo Antifungal Assays*

Stock cultures of *B. cinerea* (CECT-20973) and *P. expansum* (CECT-20906) were supplied by the Spanish Type Culture Collection (Burjassot, Valencia, Spain). These fungal strains were inoculated on to potato dextrose agar (PDA; Scharlab, Barcelona, Catalonia, Spain) in the dark and incubated at 25 °C until sporulation. Conidia were scraped from the cultures using a sterile loop and subsequently filtered and transferred to test tubes with sterile distilled water and 0.01% Tween 85. The suspensions were adjusted by means of an haemocytometer at 3 × 10^4 conidia·mL^{-1} for *B. cinerea* and 1 × 10^5 conidia·mL^{-1} for *P. expansum*, according to other studies [31,32].

Each fruit was wounded (approximately 1.6 mm in diameter and 2 mm deep) at one point of the fruit equator using the tip of a stainless-steel rod and inoculated with a micropipette with 100 μL of the correspondent spore suspension of *B. cinerea* and 20 μL in the case of *P. expansum*.

For the assessment of the coatings' curative activity, the fruit was first inoculated with the different fungal strain and after 24 h, samples were coated as previously described. Once dried, the pieces of fruit were placed on to perforated plastic trays avoiding any direct contact between them and incubated at 20 °C and 85% ± 5% RH. Twelve pieces of fruit were used per treatment. The control fruit was also inoculated using the same procedure, and afterwards immersed in water as previously described.

Disease incidence (% of infected fruit) and severity (lesion diameter) were assessed after 2, 5, 7, 9, and 12 days of incubation at 20 °C and 55% RH. The lesion diameter (mm) was evaluated by using the ImageJ 1.52a software (National Institutes of Health, Bethesda, Maryland, USA).

2.6. *Sensory Analysis*

A 40-member non-trained panel carried out the sensory evaluation of the samples 2 days after the coating application was performed. The sensory analysis was performed using whole fresh apples and, as control, uncoated fresh samples were used. All the samples, which were previously cut into wedges (1/8 of the whole apple), were presented to the judges at the same time. The judges were asked to evaluate the samples in terms of appearance and aroma, flavor, and overall preference using a 9-point hedonic scale (1 = "dislike extremely", 9 = "like extremely").

This sensory evaluation poses no hazard to human health taking into account the low level of carvacrol ingested and that carvacrol has been recognized as a food additive and as a flavoring substance by the Food and Agriculture Organization of the United Nations (FAO) and European Food Safety Authority (EFSA), respectively [33,34]. The samples were randomly presented with a three-digit code. All of the evaluations were conducted in an EU homologated sensory room.

The panelists were supplied with a rating sheet containing information on the evaluation procedure, in addition to the general oral instructions and individual clarifications as required. The panelists were also required to cleanse their palate with mineral water between the testing of different samples. The panelist's average responses were considered for each attribute.

2.7. Statistical Analysis

The statistical analyses of the results were performed through an analysis of variance (ANOVA) using Statgraphics Centurion XVI.II (StatPoint Technologies Inc., Warrenton, VA, USA). Fisher's least significant difference (LSD) test was used at the 95% confidence level to determine significant differences between means.

3. Results

3.1. Rheological Behavior of the Coating Forming Dispersions

The gravitational drainage of CFD after application occurs before the coating dries, depending on the liquid viscosity, greatly affecting the thickness of the applied coating, which, in turn, determines its barrier capacity. All the formulations exhibited Newtonian behavior in the low shear rate range considered (0–150 s^{-1}) at the solid concentrations used. The gravitational drainage occurs at low shear rates, in the order of 1–10 s^{-1} [35] and the obtained viscosity values correspond to this range. The viscosities of the CFD are shown in Table 1. The incorporation of carvacrol was observed to have no significant effect ($p > 0.05$) on the viscosity value of the pure PVA solution. This could be attributed to the changes in the PVA concentration in the continuous phase due to its interfacial adsorption at the oil-water interphase, as observed by other authors for some polymer solutions containing essential oils [36]. On the other hand, the viscosity significantly rose ($p < 0.05$) when the starch ratio or carvacrol content increased in the polymer blends (Table 1), in agreement with the thickening power of starch or the rise in the dispersed phase concentration (carvacrol). The different values in the viscosity of CFD will affect the coating thickness or surface density of the solids in the coatings.

3.2. Carvacrol Retention in the Films

The carvacrol retention in the films for each formulation is also shown in Table 1. Remarkable losses of carvacrol were expected during the film drying step due to the emulsion destabilization (droplet flocculation and creaming) that occurs in line with the water evaporation and the steam drag effect at the film surface of the creamed droplets, as reported by other authors for cast films containing carvacrol or similar volatile compounds [27,37,38]. Nevertheless, the CA-loaded films exhibited moderate retention values, of about 44%–59% depending on the formulation. The maximum retention capacity was found for the 100PVA:0S-C40 coating (Table 1) with similar values to those found by other authors for PVA films with carvacrol [39]. The retention of carvacrol was promoted by the presence of residual acetyl groups in the PVA chains. As reported by [40], the acetyl groups undergo ionization, generating negative charges in the polymer chain that can interact with the acidic phenolic group of CA. This mechanism promotes the binding of CA to the polymer chains, thus contributing to an increase in its effective retention in the polymer matrix, since the bonded carvacrol is not emulsified and so, insensitive to the emulsion destabilization and evaporation by the steam drag effect [39,41]. Nevertheless, the low viscosity of the polymer aqueous phase was a limitation for the carvacrol retention in emulsified systems during the coating drying step. In emulsified polymer systems, high values of viscosity contribute to reducing the volatile losses since high viscosity limits the creaming and surface evaporation. The relatively low viscosity of the used CFD led to moderate retention values, in comparison with other studies that considered 15% PVA and 2% CA in the aqueous dispersion [41].

The addition of starch to the CFD significantly ($p < 0.05$) decreased the carvacrol retention despite the increase in viscosity. This can be attributed to the lower degree of affinity of the starch and carvacrol compared with that of PVA-C, and the subsequent increase in the emulsified carvacrol content sensitive to destabilization processes and evaporation by means of the steam drag effect. For the same polymer composition in the formulation, the retention efficiency increased as the carvacrol content rose; this can be explained by the increase in the viscosity and the reduction in the creaming rate when the concentration of dispersed phase rose.

The FESEM micrographs of the films obtained with carvacrol (Figure 1) confirmed the abovementioned effects. In the film with only PVA, very small carvacrol droplets were observed, which indicates the PVA's stabilizing effects in carvacrol (bonded or emulsified) that limit the coalescence and creaming phenomena, increasing the carvacrol retention in the film. In PVA-starch blend films, coalescence of carvacrol can be observed forming big liquid clusters entrapped into the polymer matrix during the film forming process (small arrows in Figure 1). In the starch-PVA blend with the highest ratio of carvacrol, the top part of the film during the film-forming step and water evaporation appeared completely flooded by carvacrol, thus revealing the coalescence and creaming of the carvacrol droplets during this step. Differences in the film microstructure and carvacrol distribution could affect the carvacrol release, thus affecting its antifungal action.

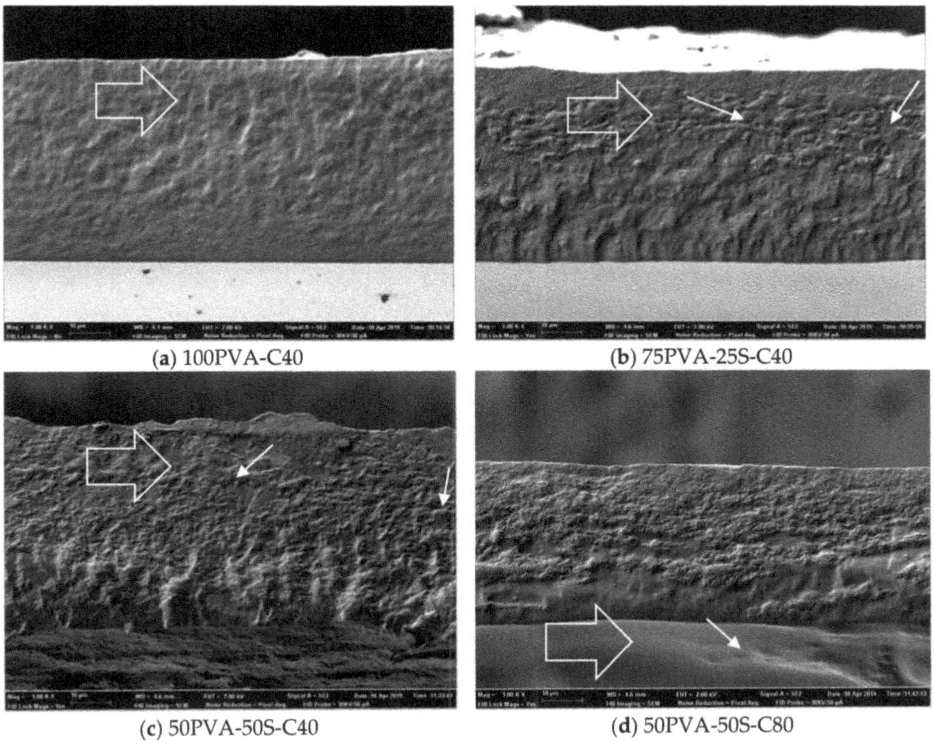

Figure 1. FESEM micrographs of the cross section of PVA composite films containing carvacrol. Big arrows indicate the upper part of the film in contact with the air during the drying process (top) and small arrows, the presence of carvacrol liquid clusters. (**a**) 100PVA-C40; (**b**) 75PVA-25S-C40; (**c**) 50PVA-50S-C40; (**d**) 50PVA-50S-C80.

3.3. *Effect of Coatings on the Postharvest Quality of Apples*

Table 2 shows the values of the surface density of solids (SDS), the rate of the relative weight loss, the fruit firmness parameters, and the respiration rates of the apple samples after two different storage times. These parameters allow for the evaluation of the relevance of the coatings on the quality changes in apple during storage.

Table 2. Surface solids density (SSD), weight loss rate, force and distance at the failure point (Fmax, dmax), O_2 consumption rate and CO_2 production rate, and respiration quotient (RQ) of coated and uncoated (control) apples.

Property	Control	100PVA:0S	100PVA:0S-C40	75PVA:25S-C40	50PVA:50S-C40	50PVA:50S-C80
SSD (g/m^2)	-	0.352 (0.112) ^{a}	0.30 (0.07) ^{a}	0.22 (0.08) ^{a}	0.49 (0.13) ^{b}	0.53 (0.09) ^{b}
Weight loss rate (day^{-1})	−0.34 (0.08) ^{a}	−0.32 (0.12) ^{a}	−0.295 (0.012) ^{a}	−0.33 (0.07) ^{a}	−0.34 (0.05) ^{a}	−0.35 (0.05) ^{a}
F_{max} (N)	31 (3) ^{a}	33 (7) ^{a}	32 (4) ^{a}	34 (4) ^{a}	32 (4) ^{a}	29 (3) ^{a}
d_{max} (mm)	3.9 (0.2) ^{bc}	4.0 (1.3) ^{bc}	4.4 (0.2) ^{c}	3.9 (0.6) ^{bc}	3.7 (0.5) ^{bc}	3.5 (0.4) ^{ab}
R_{O2} t = 3 *	11 (5) ^{ab}	6 (5) ^{a}	4.8 (1.5) ^{a}	11.9 (0.6) ^{ab}	9 (4) ^{ab}	15 (2) ^{b}
R_{CO2} t = 3 *	16 (4) ^{b}	6.6 (0.9) ^{a}	11.9 (1.4) ^{b}	13 (3) ^{ab}	12 (6) ^{ab}	14.5 (1.9) ^{b}
RQ t = 3	1.4 (0.5) ^{a}	0.8 (0.5) ^{a}	1.8 (0.3) ^{a}	1.358 (0.003) ^{a}	1.3 (0.6) ^{a}	1.26 (0.03) ^{a}
R_{O2} t = 14 *	11 (3) ^{bc}	6 (0.5) ^{a}	6.7 (1.2) ^{ab}	13 (3) ^{c}	12 (5) ^{bc}	10.5 (0.8) ^{bc}
R_{CO2} t = 14 *	16 (10) ^{bc}	3 (4) ^{a}	9.7 (1.4) ^{b}	16.2 (0.8) ^{bc}	10 (3) ^{b}	19 (2) ^{c}
RQ t = 14	1.46 (0.13) ^{bc}	1.11 (0.07) ^{a}	1.708 (0.107) ^{c}	1.308 (0.097) ^{ab}	1.435 (0.102) ^{b}	1.38 (0.07) ^{b}

*: ($mL \cdot kg^{-1} \cdot h^{-1}$). $^{a–c}$: Different superscript letters within the same row indicate significant differences ($p < 0.05$).

The coating thickness will affect the coating efficiency and its value would be related with the obtained SDS value, although the homogeneity of the solid distribution will also have a significant impact. SDS depends on the amount of coating that adheres to the surface of the fruit and the total solid content of the formulation, and it is greatly affected by the wettability, extensibility, and viscosity of the CFD [42]. The obtained SDS values ranged between 0.3–0.5 g/m^2, depending on the formulation. The amount of adhered solids was relatively low in comparison with similar studies, where thicker dispersions were used [42]. Likewise, the 50PVA:50S-C40 and 50PVA: 50S-C80 coatings presented significantly higher SDS values ($p < 0.05$) compared to the rest of the dispersions, which can be attributed to their higher viscosity values (Table 2). A high degree of viscosity limits the gravitational drainage of the applied dispersion before the drying process is applied to the coating, and so promotes the retention of a greater surface density of the CFD. Therefore, a greater coating thickness could be expected for the thickest formulations of 1:1 starch-PVA blends with carvacrol. This would also imply a greater amount of active compound on the fruit surface.

The firmness of the fruit was evaluated through the maximum force and the penetrated distance values at the failure point, since it represents the deformability of the flesh associated with differences in the cell turgor [42]. At initial time (t = 0), the uncoated control fruit exhibited a force value of 29 ± 3 N and a deformation distance of 2.9 ± 0.4 mm. After two storage weeks, the distance at failure point increased in all (coated and uncoated) of the apple samples with respect to the initial value, thus indicating the loss of cellular turgor throughout time due to the progressive dehydration of the samples at the surface level, which favors the sample deformation without break. However, no significant changes occurred in the maximum force value, which presents similar values for every sample. The samples exhibiting the greatest SDS values have the lowest deformability values, while those coated with pure PVA with carvacrol (with a low SDS value) were the most deformable. However, non-significant differences were found regarding the weight loss rate of the different samples (uncoated and coated) during the 14 storage days. This reflects a mild barrier effect of the coatings to water exchanges, which may be due to the highly hydrophilic nature of the polymers that enhanced the water vapor permeability and to the low amount of adhered solids, which reflected the very limited thickness of the coatings. An increase in the polymer concentration in the CFD, and so in viscosity, could enhance the coating thickness and boost the barrier effect to water exchanges. However, the industrial application of the CFD points to the need for low viscosity in order to facilitate their manipulation.

The effect of the coatings on the respiration rate of apples (uncoated and coated samples) was evaluated through O_2 consumption and CO_2 generation and the respiration quotient (RQ) after 3 and 14 days of storage (Table 2). In uncoated apples (control), these values were 11 ± 5 and 16 ± 4 $mL \cdot kg^{-1} \cdot h^{-1}$ for O_2 consumption and CO_2 generation, respectively, coinciding with other studies [42]. Coatings can serve as gas barriers, which could reduce respiration rates [43], due to a blockage of the surface pores. A lower respiration rate is associated with a lower exchange of gases and, therefore, a more limited availability of oxygen to allow respiration. However, changes in the internal atmosphere of the fruit depend on the type of material used, the homogeneity and thickness of the coating, and potential interactions with the natural wax of the fruits, etc. [42,44]. In general, a decrease in the O_2 consumption and CO_2 production was observed in the coated samples, although the great variability found made this decrease non-significant in most cases, despite the fact that these hydrophilic coatings represent an excellent barrier to oxygen [45]. Samples coated with pure PVA, containing or not carvacrol, exhibited the lowest respiration rates in line with the lower oxygen permeability of PVA [22]. The variability in the values can be attributed to the natural variability in fruits and to the small amount of coating material deposited on the surface that may lead to the incomplete coverage of the fruit once the coating dries [46]. In general, no significant changes in the respiration patterns were observed throughout the storage period.

The respiration quotient (RQ) ranges from 0.7–1.3 in aerobic respiration depending on the metabolic substrate [47]. In general, this respiratory quotient did not change over time, which shows that there were no changes in the metabolic pathways during the time of analysis. Nor were any significant differences found in the RQ of the different treatments, except for the formulation with pure PVA with carvacrol, where the RQ was higher after both 3 and 14 days of storage, in line with the lowest oxygen consumption rates with normal CO_2 production rates. Carvacrol incorporation in pure PVA did not reduce the great barrier capacity of PVA to oxygen, but increased the CO_2 production, in comparison with pure PVA.

These analyses led to the conclusion that applied coatings were not effective at controlling the water vapor exchanges of the fruits which produced a slight increase in the fruit deformability during storage, although pure PVA based coatings limited the oxygen consumption to a greater extent than those of PVA-starch blends with carvacrol. This compound did not induce negative changes in the metabolic pathway of apple, as deduced from the values of the respiration quotient. Likewise, it is remarkable that neither the respiration pattern nor the symptoms observed on the fruit surface pointed to any phytotoxicity of the carvacrol in the coated fruits.

3.4. Fungal Decay of Apples

Numerous studies have previously shown that the application of coatings with the incorporation of essential oils, or some of their pure components, contributed to the control of various diseases caused by pathogens in the postharvest storage of fruit (including *B. cinerea* and *P. expansum*) [48–50]. Nevertheless, a wide variability in the efficiency of the disease control can be found due to numerous factors that influence the antifungal properties of the coatings. The nature of the coating matrix, the type and concentration of antifungal compounds used, the species and strains of the target postharvest pathogens, the cultivar and the physical and physiological conditions of the host fruit and the postharvest environmental conditions are among the most important.

The carvacrol concentrations in the coatings, taking into account its mass fraction in the CFD and assuming the same retention percentage as in the films, were: 0.17, 0.10, 0.07, and 0.14 g C/g dry coating for the 100PVA:0S-C40, 75PVA:250S-C40, 50PVA:50S-C40, and 50PVA:50S-C80 formulations, respectively. This implies samples with a higher or lower content of the active compound, whose release rate will also influence the effectiveness at controlling fungal growth when the minimal inhibitory concentration is reached at the infection point. In terms of the carvacrol release, different behaviors could be expected as a function of the carvacrol load in the coating and its distribution in the polymer matrix as a consequence of the different PVA-starch-carvacrol interactions and the final film microstructure (Figure 1). The greater chemical affinity of PVA with carvacrol could limit its release from the matrix when it is richer in PVA. Likewise, the presence of starch in coating formulations could partially inhibit the action of the antimicrobial agent due to its nutritional effect that can favor the growth of fungi, such as *Botrytis* [42]. All of these aspects can contribute to the different growth inhibition behavior observed for the coatings in each fungus.

Figure 2 shows the incidence level of both fungi as a function of storage time both for the control samples and those coated with the different formulations. The incidence percentage reached a practically constant value from 5–8 storage days onwards, depending on the treatment, which was affected by the type of coating and fungus. A 100% incidence of *Botrytis* was observed for both the control samples and those coated with pure PVA without carvacrol, whereas lower percentages (67%–83%) were reached in samples coated with formulations containing carvacrol. In the case of *Penicillium*, the asymptotic incidence level was lower than in *Botrytis*, being more sensitive to the different treatments. Specifically, the greatest incidence level of *Penicillium* was observed for the treatment with pure PVA without carvacrol, which could be related with the ability of *Penicillium* to use PVA as a carbon source for growth purposes [51,52]. On the other hand, the lowest incidence level was detected for the treatments with starch-PVA blends with carvacrol and the intermediate values for both the control samples and those coated with pure PVA with carvacrol. These results indicate the different sensitivity of each fungus to the coating action and carvacrol effect. Whereas every coating with carvacrol reduced the incidence level of *Botrytis*, the incidence of *Penicillium* was only notably affected by the coatings formed by PVA-starch blends, where a faster release of the active compound could be expected due to its weaker bonding to the polymer matrix.

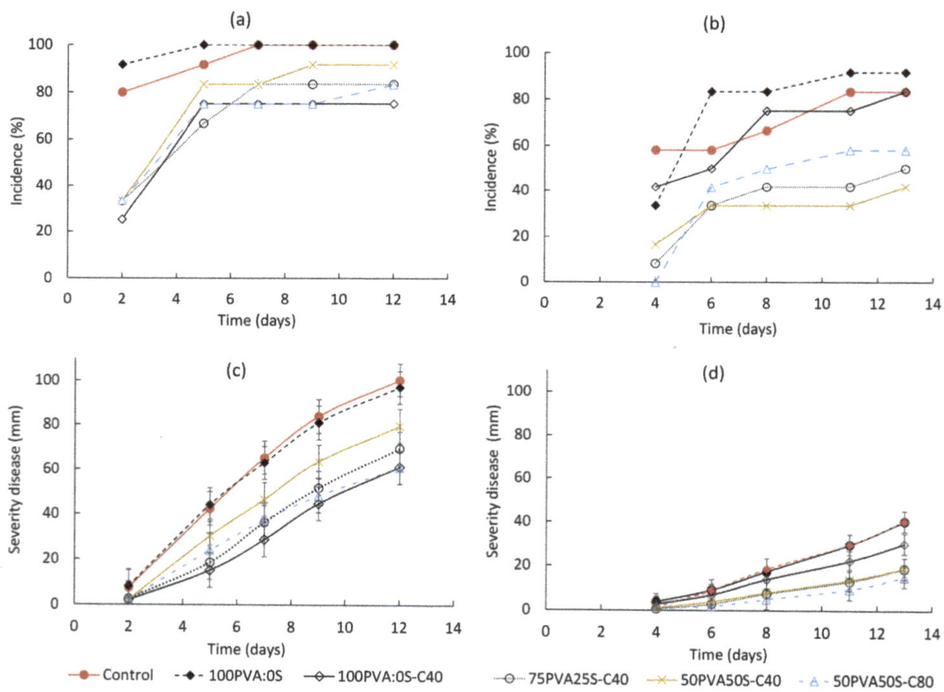

Figure 2. Incidence (%) and severity disease (mm) in apple inoculated with (**a**,**c**) *Botrytis cinerea* and (**b**,**d**) *Penicillium expansum* throughout the incubation time.

Figure 2 also shows the values of the diameter of the lesions (mm), which represent the severity of the disease caused by both fungi (*B. cinerea* and *P. expansum*) in the fruit throughout the storage period. The MANOVA analysis showed that both factors, coating treatment and storage time, significantly affected ($p < 0.05$) the disease's severity, without there being any significant interactions. The disease's severity increased throughout storage time in every case, more slowly in the samples coated with films containing carvacrol, without any significant differences between the control samples and those coated with PVA without carvacrol. The values of disease severity observed for *Penicillium* were a great deal lower than for *Botrytis*, with different trends concerning the effect of the carvacrol content. In *Penicillium*, the coatings with pure PVA and carvacrol reduced the growth of the lesion diameters to a lower extent than the blend coatings of starch and PVA with carvacrol. In contrast, the reduction in the growth of the fungal lesion for *Botrytis* was less significant with the 50PVA50S-CA40 coating, which contained the lowest final concentration of carvacrol. Therefore, both the concentration of the active compound and its release rate from the polymer matrix affected the fungal growth, depending on the fungal physiology and its sensitivity to the active compound in the different growth stages.

The antifungal activity of carvacrol has been related to the severe damage to the fungal membranes and cell walls, which led to the morphological deformations, collapse and deterioration of the conidia, and/or hyphae [19]. Antifungal effects were also observed by other authors for thymol essential oil (EO) in *Red Fuji* apple [53]. These authors showed that, additionally to the known antifungal activity of thymol essential oil (with thymol and carvacrol as major components), the efficacy of thyme essential oil was also related to the induction of host resistance, since the state of alertness is activated in the fruit after the oil application. Nevertheless, these authors also pointed out that the direct application of the EO provoked the appearance of a certain degree of phytotoxicity in the samples, which was not observed in the samples submitted to the studied treatments. However, the progressive volatilization of carvacrol may reduce its antifungal action throughout the storage time.

To summarize, the application of carvacrol-loaded coatings exerted a positive antifungal effect on apples, as significant reductions in the severity and incidence level of both *P. expansum* and *B. cinnerea* were observed. Thus, after 12 storage days, the severity of the damage was reduced by around 30%, and even by up to 33%, with respect to the control samples when using carvacrol-loaded coatings for *Botrytis* and *Penicillium*, respectively. Likewise, in the same period of time, the incidence level decreased by up to 27% and by around 40% for *Botrytis* and *Penicillium*, respectively.

3.5. Sensory Evaluation

The results of the sensory analysis carried out with the formulations applied to whole apples are summarized in Figure 3. As can be observed, the appearance of the apple was not significantly affected by the coating applications. On the contrary, the judges found significant differences ($p < 0.05$) regarding the aroma and flavor attributes between uncoated samples or ones coated only with pure PVA and those samples treated with the active coatings due to the negative impact of the odor and taste of carvacrol. Similar results have been previously observed by other authors for other carvacrol-loaded coatings [54].

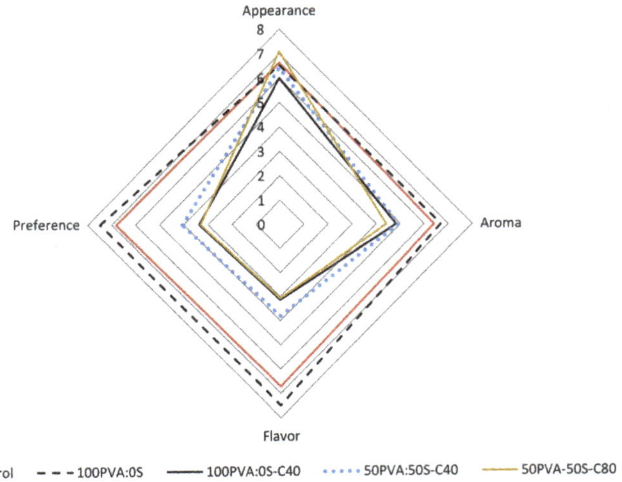

Figure 3. Sensory profile of the uncoated (control) and the different coating formulations applied onto whole fresh apples in terms of appearance, aroma, flavor, and preference.

4. Conclusions

None of the coating formulations reduced the weight loss or promoted significant changes in the respiration rates in apples, probably due to the low surface solid density of these coatings. Despite that, the application of carvacrol-loaded coatings was effective at reducing the incidence and severity of the black and blue molds caused by *B. cinnerea* and *P. expansum*. An analysis of the impact of the carvacrol-loaded coating on the apple sensory attributes revealed that the threshold of unpleasantness of aroma and flavor perception was reached when the coatings were applied and the aroma and flavor of the coated apples were negatively affected. Therefore, the application of these kinds of coatings to apples is recommended only as a pre-harvest treatment because of the high antifungal efficiency and low degree of fitotoxicity. Further studies are needed to evaluate if the organoleptic properties of the product change when those coatings are applied as a pre-harvest treatment.

Author Contributions: Conceptualization, C.G.M. and A.C.; methodology, M.S. and M.E.M.-E.; formal analysis, M.S.; writing—original draft preparation, M.S. and M.E.M.-E.; writing—review and editing, C.G.M. and A.C.; supervision, C.G.M.; funding acquisition, A.C. and C.G.M. All authors have read and agreed to the published version of the manuscript.

Funding: This research was funded by the Agencia Estatal de Investigación (Spain) through the projects RTA2015-00037-C02-00 and PID2019-105207RB-I00.

Conflicts of Interest: The authors declare no conflict of interest.

References

1. Gong, D.; Bi, Y.; Jiang, H.; Xue, S.; Wang, Z.; Li, Y.; Zong, Y.; Prusky, D. A comparison of postharvest physiology, quality and volatile compounds of 'Fuji' and 'Delicious' apples inoculated with Penicillium expansum. *Postharvest Biol. Technol.* **2019**, *150*, 95–104. [CrossRef]
2. Ma, L.; He, J.; Liu, H.; Zhou, H. The phenylpropanoid pathway affects apple fruit resistance to botrytis cinerea. *J. Phytopathol.* **2017**, *166*, 206–215. [CrossRef]
3. Nikkhah, M.; Hashemi, M.; Najafi, M.B.H.; Farhoosh, R. Synergistic effects of some essential oils against fungal spoilage on pear fruit. *Int. J. Food Microbiol.* **2017**, *257*, 285–294. [CrossRef] [PubMed]
4. Batta, Y. Postharvest biological control of apple gray mold by Trichoderma harzianum Rifai formulated in an invert emulsion. *Crop. Prot.* **2004**, *23*, 19–26. [CrossRef]
5. Neto, A.C.D.R.; Navarro, B.B.; Canton, L.; Maraschin, M.; Di Piero, R.M. Antifungal activity of palmarosa (Cymbopogon martinii), tea tree (Melaleuca alternifolia) and star anise (Illicium verum) essential oils against Penicillium expansum and their mechanisms of action. *LWT* **2019**, *105*, 385–392. [CrossRef]
6. Romanazzi, G.; Feliziani, E. Botrytis cinerea (Gray Mold). In *Postharvest Decay*; Elsevier BV: Amsterdam, The Netherlands, 2014; pp. 131–146.
7. Dhall, R.K. Advances in edible coatings for fresh fruits and vegetables: A review. *Crit. Rev. Food Sci. Nutr.* **2013**, *53*, 435–450. [CrossRef] [PubMed]
8. Lin, D.; Zhao, Y. Innovations in the development and application of edible coatings for fresh and minimally processed fruits and vegetables. *Compr. Rev. Food Sci. Food Saf.* **2007**, *6*, 60–75. [CrossRef]
9. Sánchez-González, L.; Vargas, M.; González-Martínez, C.; Chiralt, A.; Cháfer, M. Use of essential oils in bioactive edible coatings: A review. *Food Eng. Rev.* **2011**, *3*, 1–16. [CrossRef]
10. Burt, S. Essential oils: Their antibacterial properties and potential applications in foods—A review. *Int. J. Food Microbiol.* **2004**, *94*, 223–253. [CrossRef]
11. Combrinck, S.; Regnier, T.; Kamatou, G. In vitro activity of eighteen essential oils and some major components against common postharvest fungal pathogens of fruit. *Ind. Crop. Prod.* **2011**, *33*, 344–349. [CrossRef]
12. Prakash, B.; Kedia, A.; Mishra, P.K.; Dubey, N. Plant essential oils as food preservatives to control moulds, mycotoxin contamination and oxidative deterioration of agri-food commodities—Potentials and challenges. *Food Control* **2015**, *47*, 381–391. [CrossRef]
13. Sivakumar, D.; Bautista-Baños, S. A review on the use of essential oils for postharvest decay control and maintenance of fruit quality during storage. *Crop. Prot.* **2014**, *64*, 27–37. [CrossRef]
14. Ramos, M.; Jiménez, A.; Garrigós, M.C. Carvacrol-Based Films. In *Antimicrobial Food Packaging*; Elsevier BV: Amsterdam, The Netherlands, 2016; pp. 329–338.

15. Abbaszadeh, S.; Sharifzadeh, A.; Shokri, H.; Khosravi, A. Antifungal efficacy of thymol, carvacrol, eugenol and menthol as alternative agents to control the growth of food-relevant fungi. *J. Mycol. Médicale* **2014**, *24*, e51–e56. [CrossRef] [PubMed]
16. Camele, I.; Altieri, L.; De Martino, L.; De Feo, V.; Mancini, E.; Rana, G.L. In vitro control of post-harvest fruit rot fungi by some plant essential oil components. *Int. J. Mol. Sci.* **2012**, *13*, 2290–2300. [CrossRef] [PubMed]
17. De Souza, E.L.; Sales, C.V.; De Oliveira, C.E.V.; Lopes, L.A.A.; Da Conceição, M.L.; Berger, L.R.R.; Stamford, T.C.M. Efficacy of a coating composed of chitosan from Mucor circinelloides and carvacrol to control Aspergillus flavus and the quality of cherry tomato fruits. *Front. Microbiol.* **2015**, *6*, 732. [CrossRef] [PubMed]
18. Saad, I.K.; Hassan, B.; Soumya, E.; Moulay, S.; Mounyr, B. Antifungal activity and physico-chemical surface properties of the momentaneously exposed penicillium expansum spores to carvacrol. *Res. J. Microbiol.* **2016**, *11*, 178–185. [CrossRef]
19. Neri, F.; Mari, M.; Brigati, S. Control of Penicillium expansum by plant volatile compounds. *Plant Pathol.* **2006**, *55*, 100–105. [CrossRef]
20. Zabka, M.; Pavela, R. Antifungal efficacy of some natural phenolic compounds against significant pathogenic and toxinogenic filamentous fungi. *Chemosphere* **2013**, *93*, 1051–1056. [CrossRef]
21. Sapper, M.; Chiralt, A. Starch-based coatings for preservation of fruits and vegetables. *Coatings* **2018**, *8*, 152. [CrossRef]
22. Cano, A.I.; Cháfer, M.; Chiralt, A.; González-Martínez, C. Physical and microstructural properties of biodegradable films based on pea starch and PVA. *J. Food Eng.* **2015**, *167*, 59–64. [CrossRef]
23. Jayakumar, A.; Heera, K.V.; Sumi, T.S.; Joseph, M.; Mathew, S.; Praveen, G.; Nair, I.C.; Radhakrishnan, E.K. Starch-PVA composite films with zinc-oxide nanoparticles and phytochemicals as intelligent pH sensing wraps for food packaging application. *Int. J. Biol. Macromol.* **2019**, *136*, 395–403. [CrossRef] [PubMed]
24. Priya, B.; Gupta, V.K.; Pathania, D.; Singha, A.S. Synthesis, characterization and antibacterial activity of biodegradable starch/PVA composite films reinforced with cellulosic fibre. *Carbohydr. Polym.* **2014**, *109*, 171–179. [CrossRef] [PubMed]
25. Russo, M.A.; O'Sullivan, C.; Rounsefell, B.; Halley, P.; Truss, R.; Clarke, W.P. The anaerobic degradability of thermoplastic starch: Polyvinyl alcohol blends: Potential biodegradable food packaging materials. *Bioresour. Technol.* **2009**, *100*, 1705–1710. [CrossRef] [PubMed]
26. He, L.; Lan, W.; Ahmed, S.; Qin, W.; Liu, Y. Electrospun polyvinyl alcohol film containing pomegranate peel extract and sodium dehydroacetate for use as food packaging. *Food Packag. Shelf Life* **2019**, *22*, 100390. [CrossRef]
27. Tampau, A.; González-Martinez, C.; Chiralt, A. Carvacrol encapsulation in starch or PCL based matrices by electrospinning. *J. Food Eng.* **2017**, *214*, 245–256. [CrossRef]
28. Marín, A.; Atarés, L.; Cháfer, M.; Chiralt, A. Properties of biopolymer dispersions and films used as carriers of the biocontrol agent Candida sake CPA-1. *LWT* **2017**, *79*, 60–69. [CrossRef]
29. Castelló, M.; Fito, P.; Chiralt, A. Changes in respiration rate and physical properties of strawberries due to osmotic dehydration and storage. *J. Food Eng.* **2010**, *97*, 64–71. [CrossRef]
30. Saei, A.; Tustin, D.; Zamani, Z.; Talaie, A.; Hall, A. Cropping effects on the loss of apple fruit firmness during storage: The relationship between texture retention and fruit dry matter concentration. *Sci. Hortic.* **2011**, *130*, 256–265. [CrossRef]
31. Baert, K.; Devlieghere, F.; Bo, L.; Debevere, J.; De Meulenaer, B. The effect of inoculum size on the growth of Penicillium expansum in apples. *Food Microbiol.* **2008**, *25*, 212–217. [CrossRef]
32. Daniel, C.K.; Lennox, C.L.; Vries, F.A. In vivo application of garlic extracts in combination with clove oil to prevent postharvest decay caused by Botrytis cinerea, Penicillium expansum and Neofabraea alba on apples. *Postharvest Biol. Technol.* **2015**, *99*, 88–92. [CrossRef]
33. FAO; WHO. Expert Committe on Food Additives Fitthy-Fifth Report. 2001. Available online: http://apps.who.int/iris/bitstream/10665/42388/1/WHO_TRS:901.pdf (accessed on 4 August 2017).
34. EFSA. Scientific Opinion on the Safety and efficacy of phenol derivatives containing ring-alkyl, ring-alkoxy and side-chains with an oxygenated functional group (chemical group 25) when used as flavourings for all species. *Eur. Food Saf. Auth. J.* **2012**, *10*, 2573. [CrossRef]
35. Steffe, J.F. *Rheological Methods in Food Process Engineering*, 2nd ed.; Freeman Press: East Lansing, MI, USA, 1996.

36. Sánchez-González, L.; Cháfer, M.; Chiralt, A.; González-Martínez, C. Physical properties of edible chitosan films containing bergamot essential oil and their inhibitory action on Penicillium italicum. *Carbohydr. Polym.* **2010**, *82*, 277–283. [CrossRef]
37. Perdones, Á.; Escriche, I.; Chiralt, A.; Vargas, M. Effect of chitosan-lemon essential oil coatings on volatile profile of strawberries during storage. *Food Chem.* **2016**, *197*, 979–986. [CrossRef]
38. Talón, E.; Vargas, M.; Chiralt, A.; González-Martínez, C. Antioxidant starch-based films with encapsulated eugenol. Application to sunflower oil preservation. *LWT* **2019**, *113*, 108290. [CrossRef]
39. Andrade, J.; González-Martínez, C.; Chiralt, A. The Incorporation of Carvacrol into Poly (vinyl alcohol) Films Encapsulated in Lecithin Liposomes. *Polymer* **2020**, *12*, 497. [CrossRef] [PubMed]
40. Wiśniewska, M.; Bogatyrov, V.M.; Ostolska, I.; Szewczuk-Karpisz, K.; Terpiłowski, K.; Nosal-Wiercińska, A. Impact of poly(vinyl alcohol) adsorption on the surface characteristics of mixed oxide Mn_xO_y-SiO_2. *Adsorption* **2015**, *22*, 417–423. [CrossRef]
41. Tampau, A. Carvacrol Encapsulation by Electrospinning or Solvent Casting to Obtain Biodegradable Multilayer Active Films for Food Packaging Applications. Ph.D. Thesis, Universitat Politecnica de Valencia, Valencia, Spain, 2020.
42. Sapper, M.; Palou, L.; Pérez-Gago, M.B.; Chiralt, A. Antifungal starch–gellan edible coatings with thyme essential oil for the postharvest preservation of apple and persimmon. *Coatings* **2019**, *9*, 333. [CrossRef]
43. Conforti, F.D.; Totty, J.A. Effect of three lipid/hydrocolloid coatings on shelf life stability of Golden Delicious apples. *Int. J. Food Sci. Technol.* **2007**, *42*, 1101–1106. [CrossRef]
44. Smith, S.; Geeson, J.; Stow, J. Production of modified atmospheres in deciduous fruits by the use of films and coatings. *HortScience* **1987**, *22*, 772.
45. Miller, K.; Krochta, J. Oxygen and aroma barrier properties of edible films: A review. *Trends Food Sci. Technol.* **1997**, *8*, 228–237. [CrossRef]
46. Banks, N.H.; Dadzie, B.K.; Cleland, D.J. Reducing gas exchange of fruits with surface coatings. *Postharvest Biol. Technol.* **1993**, *3*, 269–284. [CrossRef]
47. Kader, A.A.; Zagory, D.; Kerbel, E.L.; Wang, C.Y. Modified atmosphere packaging of fruits and vegetables. *Crit. Rev. Food Sci. Nutr.* **1989**, *28*, 1–30. [CrossRef] [PubMed]
48. Campos-Requena, V.H.; Rivas, B.L.; Pérez, M.A.; Figueroa, C.R.; Figueroa, N.E.; Sanfuentes, E.A. Thermoplastic starch/clay nanocomposites loaded with essential oil constituents as packaging for strawberries—In Vivo antimicrobial synergy over Botrytis cinerea. *Postharvest Biol. Technol.* **2017**, *129*, 29–36. [CrossRef]
49. Tovar, C.D.G.; Chaves-López, C.; Serio, A.; Rossi, C.; Paparella, A. Chitosan coatings enriched with essential oils: Effects on fungi involved in fruit decay and mechanisms of action. *Trends Food Sci. Technol.* **2018**, *78*, 61–71. [CrossRef]
50. Perdones, A.; Sanchezgonzalez, L.; Chiralt, A.; Vargas, M. Effect of chitosan–lemon essential oil coatings on storage-keeping quality of strawberry. *Postharvest Biol. Technol.* **2012**, *70*, 32–41. [CrossRef]
51. Qian, D.; Du, G.; Chen, J. Isolation and culture characterization of a new polyvinyl alcohol-degrading strain: Penicillium sp. WSH02-21. *World J. Microbiol. Biotechnol.* **2004**, *20*, 587–591. [CrossRef]
52. Kawai, F.; Hu, X. Biochemistry of microbial polyvinyl alcohol degradation. *Appl. Microbiol. Biotechnol.* **2009**, *84*, 227–237. [CrossRef]
53. Banani, H.; Olivieri, L.; Santoro, K.; Garibaldi, A.; Gullino, M.L.; Spadaro, D. Thyme and savory essential oil efficacy and induction of resistance against botrytis cinerea through priming of defense responses in apple. *Foods* **2018**, *7*, 11. [CrossRef]
54. Cano, A.; Nácher, M.C.; Chiralt, A.; Pons, M.P.M.; Llopis, M.B.; Martínez, M.C.B.; Martínez, C.G. Quality of goat's milk cheese as affected by coating with edible chitosan-essential oil films. *Int. J. Dairy Technol.* **2016**, *70*, 68–76. [CrossRef]

Publisher's Note: MDPI stays neutral with regard to jurisdictional claims in published maps and institutional affiliations.

coatings

MDPI

Article

Controlling *Fusarium oxysporum* Tomato Fruit Rot under Tropical Condition Using Both Chitosan and Vanillin

Zahir Shah Safari, Phebe Ding *, Jaafar Juju Nakasha and Siti Fairuz Yusoff

Department of Crop Science, Faculty of Agriculture, Universiti Putra Malaysia, Serdang 43400, Malaysia; zahirshah.safari@gmail.com (Z.S.S.); jujunakasha@upm.edu.my (J.J.N.); yuezyusoff@gmail.com (S.F.Y.)
* Correspondence: phebe@upm.edu.my

Abstract: Tomato *Lycopersicon esculentum* Mill. is one of the most cultivated and widely consumed vegetables in the world. However, it is very susceptible to the infection initiated by *Fusarium oxysporum* fruit rot, which shortens post-harvest life and thus reduces market value. This disease can be regulated appropriately by the application of synthetic fungicides. However, chemical fungicides constitute a serious health risk, and have harmful environment effects and increase disease resistance, even when microbes are dead. Hence, to overcome this problem, chitosan and vanillin, which have antimicrobial bioactive properties against the growth of microorganisms, could be an alternative to disease control, while maintaining fruit quality and prolonging shelf life. The aim of this research was to evaluate the antimicrobial activity of chitosan and vanillin towards the inoculate pathogen and to investigate the effect of chitosan and vanillin coating in vivo on *Fusarium oxysporum* fruit rot and defense-related enzymes (PAL, PPO and POD). Chitosan and vanillin in aqueous solutions, i.e., 0.5% chitosan + 10 mM vanillin, 1% chitosan + 10 mM vanillin, 1.5% chitosan + 10 mM vanillin, 0.5% chitosan + 15 mM vanillin, 1% chitosan + 15 mM vanillin and 1.5% chitosan + 15 mM vanillin, were used as edible coatings on tomatoes stored at 26 ± 2 °C and 60 ± 5 relative humidity. The result revealed 1.5% chitosan + 15 mM vanillin was able to control disease incidence by 70.84% and severity by 70%. These combinations of coatings were also able to retain phenylalanine ammonia-lyase (PAL), peroxidase activity (POD), and polyphenol oxidase (PPO) enzyme activities as well as prolong shelf life of tomatoes up to 15 days.

Keywords: postharvest disease; antioxidant activity; postharvest losses; protein; phenylalanine ammonia-lyase (PAL); peroxidase activity (POD); polyphenol oxidases (PPO); Fusarium fruit rot

Citation: Safari, Z.S.; Ding, P.; Nakasha, J.J.; Yusoff, S.F. Controlling *Fusarium oxysporum* Tomato Fruit Rot under Tropical Condition Using Both Chitosan and Vanillin. *Coatings* **2021**, *11*, 367. https://doi.org/10.3390/coatings11030367

Academic Editor: Stefano Farris

Received: 10 February 2021
Accepted: 17 March 2021
Published: 23 March 2021

Publisher's Note: MDPI stays neutral with regard to jurisdictional claims in published maps and institutional affiliations.

1. Introduction

Tomato fruit (*Solanum lycopersicum* Mill.) is the second most important vegetable after the potato, and it can be consumed either raw or cooked [1]. Tomato fruit contains impressive health benefits due to its significant bioactive antioxidant compounds with β-carotene, lycopene, flavonoids, organic acids, chlorophyll and vitamins [2]. In addition, it is also low in fat and calories. Tomato fruit is a very delicate fruit vegetable that is susceptible to high postharvest losses, which vary from country to country. The postharvest loss of tomato fruit was 17.26% in India, 12% in the US, 13.75% in Egypt and 26% in Bangladesh [3,4]. One of the prominent losses was due to pathological damage that affects the quality and nutritional value of this fruit [5]. Moreover, pathological damage causes mycotoxin contamination and market value reduction of the fresh produce. Fusarium fruit rot caused by *Fusarium oxysporum* is a common disease in tomato fruit and is capable of causing fruit spoilage either in the field or in storage [6]. The application of synthetic fungicides such as carbendazim, chlorothalonil and mancozeb [7] is commonly used to control this disease. However, this control measure has negative impacts on human health and the environment due to chemical residues and an increase in pathogen resistance [8]. Furthermore, public anxiety over sustainable food production and safety has resulted in attempts to find new control agents for postharvest diseases; this has led us to study the effects of combining chitosan and vanillin as natural antimicrobials in controlling tomato postharvest disease.

Chitosan (poly-β-(1-4)-N-acetyl-D-glucosamine) and its derivative have been reported to be a potent and biodegradable alternative to synthetic fungicides [9]. Earlier studies indicated that chitosan was effective in controlling postharvest decay of many fruits and vegetables, including pomegranates, tomato fruit, strawberries, potatoes, table grapes, apples and peaches [10,11]. In addition, vanillin (4-hydroxy-3-methoxy benzaldehyde), a phenolic aldehyde organic compound derived from the vanilla bean [12], has been proven to control *Alternaria alternata*, which causes *Alternaria* rot disease in table grapes during storage at 4 ± 2 °C [8]. Temperature management during storage is also an important factor in disease management, fruit quality and the extension of shelf life. However, in developing countries, most growers and retailers store tomatoes at ambient conditions (26 ± 2 °C) due to a lack of cool storage facilities [13]. Nevertheless the effects of a coating prepared by combining chitosan, which has quality maintenance properties, and vanillin, which has antimicrobial properties, in controlling Fusarium fruit rot and postharvest quality of tomatoes at 26 ± 2 °C has yet to be studied. Therefore, this study was conducted to determine the potential of chitosan in combination with vanillin as a coating in controlling tomato fruit Fusarium fruit rot.

2. Materials and Methods

2.1. Fruit Materials

Pink color tomato fruit (10% to 30% of the surface is yellow to pink, according to USDA class 3 color) from the Syngenta 1039 variety were obtained from Weng Seng Vegetable Products Sdn. Bhd., Pahang, Malaysia. On the same day of harvesting, tomato fruit was sent to the Laboratory of Postharvest, Department of Crop Science, Faculty of Agriculture, Universiti Putra Malaysia. Fruit free from any defects and diseases, with uniform shape, maturity, color and weight and ranging between 90–110 g was used in this study.

2.2. Pathogen Inoculation

Fusarium oxysporum (MT012284) were originally isolated from tomato fruit showing fruit rot symptoms; the outer surface of the infected fruit appeared as a pale white lesion, with powdery discolored spots covered by white and pinkish mycelium. The infected area was softer and slightly sunken as compared to unaffected fruit parts (Figure 1).

Figure 1. Symptoms of tomato fruit rot: soft and sunken tissue covered by white pinkish mycelium. **A**—20% severity, **B**—40% severity, and **C**—more than 75% severity.

Fusarium oxysporum were identified based on their morphological cultural traits on fungus colony culture (conidia shape), and morphological identification was confirmed by molecular identification. Fungal DNA was extracted from freshly collected mycelium of 7-day-old cultures using the DNA extraction Kit (QIAGEN DNA Mini Kit, HB-1166, Hilden, Germany). The internal transcribed spacer (ITS) region of rDNA was amplified using the universal primers ITS1 and ITS4 (Kulatunga et al., 2016). The forward primer ITS1 (5′-TCCGTAGGTGAACCTGCGG-3′) and reverse primer ITS4 (5′-TCCTCCGCTTATTGATATGC-3′) were synthesized by First Base Laboratories Sdn Bhd, Malaysia. The ITS sequence was compared with the NCBI GenBank database (www.ncbi.nlm.nih.gov/blast, accessed on 17 May 2019) sequences using the BLAST search tool. The analyses supported the results obtained in the morphological study and thus confirmed that *Fusarium oxysporum* was identified as the causative agent of tomato fruit rot.

Tomato fruit rot causative pathogen *Fusarium oxysporium* was cultured and incubated for 7 days at 26 ± 2 °C and 60 ± 5% RH. Sterilized distilled water of 10 mL was poured into 1-week-old *Fusarium oxysporium* culture, and the surface was scraped lightly with a bent glass rod. The obtained conidial suspension was filtered over a double-layer sterilized muslin cloth and centrifuged for 5 min at 4000× *g*. The conidial counts were adjusted to 2×10^6 conidia per mL using a hemocytometer. The surface of tomato fruit was sterilized for 3 min in 0.05% sodium hypochlorite, then washed using tap water and air-dried under sterile conditions for 2 h. After drying, the fruit was dipped into antagonistic conidial suspension for 1 min, and then the fruit was allowed to dry at 26 ± 2 °C and 60 ± 5% RH for 2 h.

2.3. Preparation of Coating Solutions

The chitosan originated from shrimp-shell crustaceans with 85% deacetylation, purchased from Enviro Clean Energy Sdn. Bhd. Perintis Teknologi Pertanian, Malaysia. Meanwhile, an organic compound of 99% pure vanillin with the molecular formula $C_8H_8O_3$ was bought from Evergreen Engineering & Resources Sdn. Bhd., Selangor, Malaysia. Chitosan solutions with concentrations of 0.5%, 1%, and 1.5% *v*/*v* were prepared, the solution pH was adjusted to 5.6 with 1 M NaOH, and 0.1% Tween 20 (polyoxyethylene sorbitan monooleate, Sigma Aldrich) was added to improve the solution wettability. Distilled water containing 0.1% Tween 20 without chitosan served as a control. Vanillin powders were dissolved in 83 °C distilled water to obtain a 10 and 15 mM concentration solution by heating. Then, each vanillin solution was combined with the three chitosan concentration solutions to form 0.5% chitosan + 10 mM vanillin, 1% chitosan + 10 mM vanillin, 1.5% chitosan + 10 mM vanillin, 0.5% chitosan + 15 mM vanillin, 1% chitosan + 15 mM vanillin and 1.5% chitosan + 15 mM vanillin coating solutions.

2.4. Postharvest Coating Treatments

Tomato fruit with chlorinated water prepared from 0.05% sodium hypochlorite was dipped for 3 min prior to coating treatments [14]. The fruit was rinsed and air dried for 1 h and randomly divided into 8 lots. All 8 lots of fruit were dipped for 1 min in the coating solutions. The negative control consisted of fruit without inoculation, while the positive control fruit was inoculated with *Fusarium oxysporum* and then dipped in distilled water containing 0.1% Tween 20 for 1 min. All of the fruit was dried for 2 h at 26 ± 2 °C and 60 ± 5% RH. For each coating, six fruit per replicate were used. The fruit was then packed in 18 cm × 26 cm plastic bags of 0.05 mm thickness containing 18 holes 0.5 cm in diameter. These bags were placed in commercial corrugated fiberboard cartons of 30 cm × 25 cm × 15 cm. The cartons were then stored at 26 ± 2 °C and 60 ± 5% RH for 15 days. Each treatment was repeated four times and analysis was carried out at and interval of every 3 days.

2.5. Determination of Disease Incidence

The disease incidence (DI) was measured as the percentage of fruit displaying fruit rot symptoms, according to the method of Khaliq [15]. The DI was determined as the number of infested fruit showing symptoms of the disease, such as dots and rots, out of the total number of tomato fruit for each batch and storage interval. Six tomato fruit were distributed and used for DI. The percentage of disease was determined using the following formula (Equation (1)), as reported by Abebe et al. (2017):

$$\text{DI }(\%) = \frac{\sum(\text{DI level}) \times (\text{Number of tomato fruit at the DI level})}{\text{Total number of tomato fruit in the treatment} \times \text{The highest score }(5)} \times 100\% \quad (1)$$

2.6. Determination of Disease Severity

Tomato fruit disease severity (DS) was evaluated as described by Mohamed [16], with slight modification. Fruit DS was evaluated based on visible symptoms, spots, rot, and decayed areas on each fruit surface at every storage interval. For DS assessment, five DS scores were used, as shown in Table 1. Fruit with index scores of two, three, and four were considered to have no commercial and marketing value (Equation (2)).

$$\text{DS }(\%) = \frac{\sum(\text{Severity rating} \times \text{Number of tomato fruit clusters in the rating})}{\text{Total number of tomato fruit clusters assessed} \times \text{Highest DS scale}} \times 100\% \quad (2)$$

Table 1. Disease severity scores for disease assessment of tomato fruit.

Diseases Score	Description	Inference
0	No visible symptoms on fruit	No infection
1	1–25% of the area covered by slight necrotic inoculations	Mild infection
2	26–50% of the inoculated area covered by necrotic and white fungal mycelia	Moderate infection
3	51–75% of the sample is necrotic with the presence of spore mass	Severe infection
4	>76% Necrotic tissue with fungal mass; appears soft and decayed	Very severe/Devastating

2.7. Determination of Antioxidant Capacity and Activity

2.7.1. Supernatant Extraction

Tomato fruit pulp tissue extraction assays for total phenolic content (TPC) and antioxidant activities Radical scavenging activity by using (2,2-azino-bis (3-ethylbenzthiazoline-6-sulfonic acid) (ABTS), Ferric Reducing Antioxidant Power (FRAP) were extracted using methods defined by Rastegar and Si [17,18], with slight modification. Four grams of tomato fruit tissue from each replicate was frozen immediately in liquid N_2 and minced using a small ceramic kitchen pestle and mortar for 30 s. The ground tissue was dissolved in 10 mL 80% (*v*/*v* methanol analytical grade) and then transferred to a 100 mL conical flask, which was covered with aluminum foil. Subsequently, the homogenate was extracted under reduced light conditions by spinning on an orbital shaker at 180 rpm for 1 h. After shaking, the homogenate was filtered by Whatman No. 1 filter paper and transferred to a vial, which was covered with aluminum foil; the supernatant was kept at −20 °C until analyses.

2.7.2. Determination of Total Phenolic Content

The total phenolic content was estimated following the method described by Zainal [19], with some modifications. In brief, 150 µL aliquot of supernatant extract and 750 µL of 10% (*v*/*v*) Folin Ciocalteu reagent were mixed in test tubes covered with aluminum foil and incubated for 5 min in darkness. This was followed by the addition of 600 µL of 7.5% (*w*/*v*) sodium carbonate (Na_2CO_3). The mixture was then incubated in darkness for 30 min at room temperature 26 ± 2 °C and 60 ± 5% RH, before measuring the absorbance at 765 nm with a spectrophotometer (S1200, Spectrowave spectrophotometer, Cambridge, UK). The total phenolic content was expressed as milligram gallic acid equivalent (GAE) per 100 g fresh weight (FW), using gallic acid as the standard with $R^2 = 0.97$ and calculated using the following equation (Equation (3)):

$$\text{TPC mg GAE/100 g FW} = \frac{\text{TPC per mL sample} \times \text{dilution factor} \times \text{total sample volume used}}{\text{Sample weight}} \times 100\% \quad (3)$$

2.8. Antioxidant Activity and Capacity

2.8.1. ABTS (2,2-Azino-bis(3-ethylbenzthiazoline-6-sulfonic acid)

The antioxidant activity of tomato fruit was measured using 2,2-azino-bis, 3-ethylbenzthiazoline-6-sulfonic acid according to the methods of Aadesariya and Pinheiro [20,21], with slight modification. ABTS was formed by reacting 7 mM ABTS aqueous solution with 2.45 mM of potassium per sulphate at 26 ± 2 °C and 60 ± 5% RH for 16 h in the dark. This solution was diluted in ethanol (around 1:89 v/v) before the test and equilibrated at 30 °C to provide an absorbance of 0.700 ± 0.02 at 734 nm. The addition of 1 mL diluted ABTS solution in ethanol to 10 µL of sample extract was incubated at 30 °C for 6 min before absorbance. The inhibition percentage for the blank absorbance was then calculated at 734 nm. The percentage of ABTS free radical inhibition was determined using the equation below (Equation (4)):

$$\text{ABTS inhibition } (\%) = \frac{(A0 - A1)}{A0} \times 100\% \quad (4)$$

where A0 = absorbance of the control and A1 = absorbance of sample

Solution A was prepared by dissolving 8 mg ABTS in 1 mL of water to obtain 7 mM ABTS solution. Solution B was prepared by dissolving 13.2 mg potassium per sulphate in 10 mL water to obtain 2.45 mM solution. Solution A (0.5 mL) was mixed with 0.5 mL of solution B and allowed to sit in darkness at 26 ± 2 °C for 12–16 h before use. The ABTS radical cation in this form is stable for 16 h.

2.8.2. Ferric Reducing Antioxidant Power

Tomato fruit tissue's antioxidant capacity was calculated using ferric reducing antioxidant power (FRAP). The assay was carried out according to the methods of Briones and Thaipong [22,23], with slight modifications. In FRAP assay, the FRAP reagent was freshly prepared by mixing 10 mM of 2,4,6-tris(2-pyridyl)-s-triazine (TPTZ) in 40 mM HCl solution, 300 mM acetate buffer ($C_2H_3NaO_2 \cdot 3H_2O$, pH 3.6), and 20 mM ferric chloride in the ratio of 1:10:1 ($v/v/v$). An aliquot of 50 µL sample extract was added to 950 µL FRAP reagent and incubated in a water bath of 37 °C for 30 min. Absorbance was measured at 593 nm against a control that was prepared by adding 50 µL 80% methanol to 950 µL FRAP reagent. The standard curve was a linear line between 0 and 800 mM Trolox. The achieved results were expressed as µM Trolox equivalent (TE) of tomato fruit fresh weight using a standard curve with $R^2 = 0.98$. The obtained FRAP results were expressed in µM TE/g fresh weight and then calculated using the formula below (Equation (5)):

$$\text{FRAP } \mu\text{M TE/g FW} = \frac{\text{TE } \mu\text{M per mL} \times \text{dilution factor} \times \text{total sample volume used}}{\text{Sample weight}} \quad (5)$$

2.9. Determination of Defense Enzymes Activities

2.9.1. Protein Content

The extraction and analysis of protein were carried out using the combined techniques of Jumnongpon; Raseetha and Bonjoch [24–26], with minor modifications. The chemicals used to extract and evaluate enzymes were of analytical grade. A total of 0.5 g of frozen tomato fruit pulp tissue was immediately ground using a small ceramic kitchen pestle and mortar for 30 s on ice and homogenized with 1 mL ice-cold 50 mM phosphate buffer containing 1 M NaCl (pH 7.1). The mixture was centrifuged (Scan Speed 1730R, Scala

Scientific, the Netherlands) at 16,000× *g* at 4 °C for 20 min. The supernatant was then kept in an ice-water bath prior to the analysis.

The protein content of solutions derived from tomato fruit was measured using the Bradford procedure (Bradford 1976). The Bradford reagent was obtained from Bio-Rad Laboratories, Inc., Hercules, CA, USA. The reagent was prepared using distilled water in a 1:4 ratio; then, 1.2 mL of Bradford reagent was added with 120 µL protein supernatant and the mixture was briefly vortexed. The mixture was left to incubate for 30 min at room temperature, and the absorbance was read at 595 nm. The concentration of the extracted protein solutions from the bovine serum albumin standard curve (R^2 = 97) was quantified. The measurement was repeated three times. A standard curve plotting absorbance with various concentrations was obtained using bovine serum albumin (Sigma Chemicals Co., St. Louis, MO, USA) in the concentration range 25–400 µg/mL. The protein content in mg/mL was read against the standard curve and calculated using the formula specified by Wang [27] (Equation (6)):

$$\text{Protein content (mg/mL)} = \frac{\text{protein quality} \times \text{VT}}{\text{VS} \times \text{W}} \tag{6}$$

Protein quality results were collected in agreement with the standard curve; VT is the total volume of extraction, VS is the volume of solution for evaluation, and W is the weight of sample.

2.9.2. Determination of Phenylalanine Ammonia-Lyase (PAL) Enzyme Activity

The extraction for enzyme phenylalanine ammonia-lyase (PAL) was carried out according to Mohammed and Han [16,28], with some modifications. A total of 50 mg of frozen tissue was ground in 2 mL cold 25 mM sodium borate buffer (pH 8.8) containing 2 mM β-mercaptoethanol and 0.5 g polyvinylpyrrolidone. The homogenate was centrifuged (Scan Speed 1730R, Scala Scientific, Ede, The Netherlands) for 20 min at 16,000× *g* at 4 °C, and the supernatant was used as an enzyme source to determine the PAL activity.

PAL activity was determined by the production of cinnamate at 37 °C for 1 h; the absorbance was measured at 290 nm [29]. The assay mixture comprised 1 mL of enzyme extract and 2 mL of 50 mM sodium borate buffer (pH 8.8). The reaction started with 1 mL of 20 mM L-phenylalanine added and incubated at 37 °C for 1 h. Then, the reaction was stopped by adding 1 mL of 1 M HCl. The blank assay was performed with a mixture containing L-phenylalanine at zero incubation time. One unit of PAL activity was defined as the amount of enzyme that produced an absorbance increase of 0.01 at 290 nm per h [29]. The specific activity of the PAL enzyme was expressed as U/mg protein, where one unit of enzyme activity was defined as the production of cinnamic acid and the increase of one unit in absorbance per h. The activity of the enzyme was determined using the analytical approximation as defined in the following equation (Equation (7)):

$$\text{Unit enzyme activity (U/mL)} = \Delta\text{A 270 nm/min Test} - \Delta\text{A 270 nm/min Blank} \times 3 \times \text{df}/19.73 \times 0.1 \tag{7}$$

3 = total sample volume (mL)
df = dilution factor (weight/volume 50 mg/2 mL = 25)
19.73 = mM extinction coefficient of trans-cinnamate at 270 nm
unit definition: one unit will deaminate 1.0 µM of L-phenylalanine to trans-cinnamate and NH_3 per minute at pH 8.5 at 30 °C.

The specific activity of the enzymes was expressed in U/mg protein as followed: specific activity (U/mg protein) = unit activity (U/mL)/protein content (mg/mL) (Sigma Prod. No. P-2126).

2.9.3. Determination of Peroxidase Activity

Extraction and assay of peroxidase activity (POD) were carried out based on the combined procedure of Zhang and Raseetha [25,30], with minor modifications. A total of

0.5 g of frozen tomato fruit pulp tissue was immediately ground by using a small ceramic kitchen pestle and mortar for 30 s on ice and homogenized with 1 mL ice-cold 50 mM phosphate buffer containing 1 M NaCl (pH 7.1). The mixture was centrifuged for 20 min at 16,000× *g* at 4 °C. The supernatant was then kept in an ice-water bath prior to the analysis.

The POD activity was determined based on the development of brown coloration in the presence of H_2O_2, arising from the oxidation of guaiacol. A 20 μL sample extract supernatant was well mixed in a clean cuvette with 1.7 mL 0.1 M sodium phosphate buffer at pH 7.0 and 200 μL of 1 mM guaiacol. Then, the POD reaction was started by adding 100 μL 1.5% H_2O_2 v/v. The rate of absorbance rise at 485 nm was monitored for 3 min at 20 °C. The POD activity was expressed as U/mg protein by Kokkinakis and Ogola [31,32], as follows (Equation (8)):

$$\text{Unit activity (U/mL)} = (\Delta\text{OD/min} \times \text{V} \times \text{D})/(26.6 \times \text{d} \times \text{v}) \quad (8)$$

where ΔOD/min = the increase in absorbance at 485 nm/min, V = total amount of reaction mixture (2 mL), D = enzyme dilution factor, 26.6 = mM extinction coefficient of guaiacol at 485 nm, d = light path length (cm) and v = volume of enzyme sample (0.02 mL)

The extinction coefficient was calculated using Beer-Lambert law (ε = A/Lc): ε = extinction coefficient, A = absorption, L = path length (the thickness of the solution) and c = concentration of the solution.

The specific activity of the enzymes was expressed in U/mg protein, as follows (Equation (9)):

$$\text{Specific activity (U/mg protein)} = \text{Unit activity (U/mL)/Protein content (mg/mL)} \quad (9)$$

2.9.4. Determination of Polyphenol Oxidase Activity

Polyphenol oxidase (PPO) activity was determined based on changes in the color intensity of catechol oxidation, as described in the methods of Indunil and Mishra [33,34]. The extracted POD supernatant was used as the source of the enzyme, which was held at −20 °C. In brief, 200 μL of 0.01 M catechol was supplemented to start the reaction. The absorbance changes were recorded at 495 nm for 1 min. The PPO specific activity was determined by expressing PPO enzyme specific activity (U/mg protein) using the following equation (Equation (10)):

$$\text{Unit activity (U/mL)} = (\Delta\text{OD/min} \times \text{V} \times \text{D})/(11.3 \times \text{d} \times \text{v}) \quad (10)$$

where ΔOD/min = the change in absorbance at 485 nm/min, V = total volume of reaction mixture (2.00 mL), D = enzyme dilution factor, 11.3 = mM extinction coefficient of catechol, d = light path length (1 cm) and v = volume of enzyme sample (0.2 mL)

The extinction coefficient was calculated by the Beer-Lambert law (ε = A/Lc): ε = extinction coefficient, A = absorption, L = path length (the thickness of the solution) and c = concentration of the solution.

The specific activity of the enzymes was expressed in U/mg protein, as follows (Equation (11)):

$$\text{Specific activity (U/mg protein)} = \text{Unit activity (U/mL)/Protein content (mg/mL)} \quad (11)$$

2.10. Experimental Design and Statistical Analysis

The experiments were carried out using a completely randomized design (CRD), with eight coating treatments and four replications (Figure 2). The data obtained were analyzed using analysis of variance (ANOVA), and mean comparisons were performed using the least significant difference (LSD) at the significance level of $p \leq 0.05$. All the analyses were conducted using statistical analysis software (SAS) version 9.4 (SAS Institute Inc., Cary, NC, USA). The data in percent were transformed using a square root transformation before determining the significance level using LSD (Gomez and Gomez 1984). Pearson's coefficients correlation were conducted to correlate the determined variables. The entire experiment was repeated four times, and the data were pooled before analysis. However, the positive control fruit could no longer be used for analysis after day 12 due to high disease severity and decay.

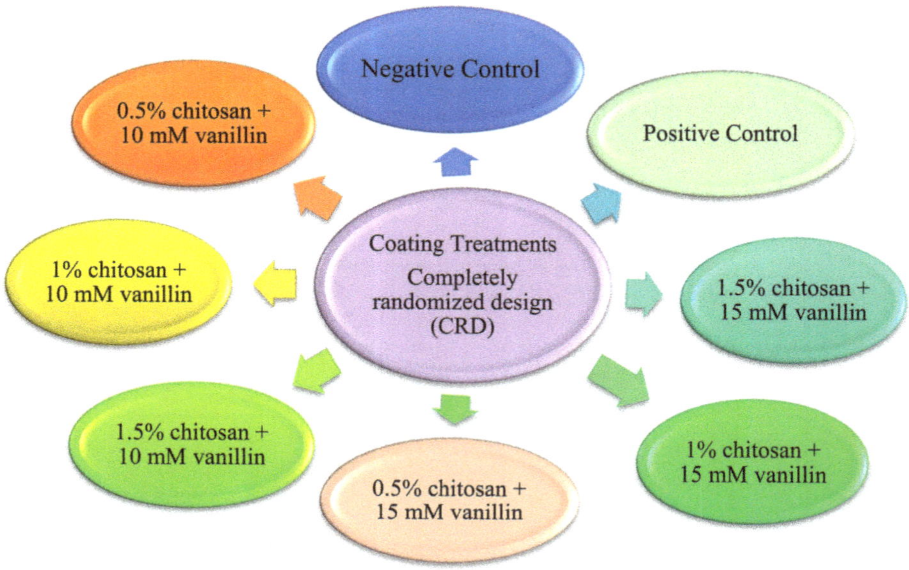

Figure 2. Schematic diagram of eight coating treatments used for *Fusarium oxysporum* inoculated tomato fruit stored for 15 days at 26 ± 2 °C and 60 ± 5% relative humidity.

3. Results

3.1. Disease Incidence and Diseases Severity

Table 2 shows that there was a significant interaction between coating treatments and storage day on the DS of tomato fruit.

Figure 3 shows that the incidence of the disease appeared after 6 days of storage, except the fruit treated with 1.5% chitosan + 15 mM vanillin. By storage day 9, DI in fruit treated with 1% chitosan + 15 mM vanillin and 1.5% chitosan + 15 mM vanillin remained low as compared to the negative and positive control fruit and those coated with 0.5% chitosan + 10 mM vanillin, 1% chitosan + 10 mM vanillin, 1.5% chitosan + 10 mM and 0.5% chitosan + 10 mM vanillin. This trend continued until storage day 15. At the end of storage, all fruit was severely infected by the disease, but the fruit treated with 1% chitosan + 15 mM vanillin and 1.5% chitosan + 15 mM vanillin showed lower incidence than other treatments.

Figure 4 shows that the DS appeared after 6 days of storage, except for the fruit treated with 1.5% chitosan + 15 mM vanillin. By storage day 9, disease severity in fruit treated with 1% chitosan + 15 mM vanillin and 1.5% chitosan + 15 mM vanillin was lower than the negative and positive control fruit and those coated with 0.5% chitosan + 10 mM vanillin, 1% chitosan + 10 mM vanillin, 1.5% chitosan + 10 mM vanillin and 0.5% chitosan + 15 mM vanillin. This trend continued until 15 days of storage. At the end of storage, all fruit were severely infected by the disease, but the fruit treated with 1% chitosan + 15 mM vanillin and 1.5% chitosan + 15 mM vanillin showed lower DS than other treatments.

From the Pearson's correlation analysis, there was strong significant positive correlation between disease incidence and severity (r = 0.94) (Table 3).

Table 2. Main and interaction effects of different coating treatments and storage days on disease incidence and severity of *Fusarium oxysporum* inoculated tomato fruit stored at 26 ± 2 °C and 60 ± 5% relative humidity for 15 days.

Factor	Disease Incidence (%)	Disease Severity (%)
Treatment	-	-
Negative control	37.50 ab [z]	40.83 ab
Positive control	44.44 a	50.83 a
0.5% chitosan + 10 mM vanillin	38.19 ab	38.33 b
1% chitosan + 10 mM vanillin	31.25 b	26.66 cd
1.5% chitosan + 10 mM vanillin	32.63 b	22.50 d
0.5% chitosan + 15 mM vanillin	31.25 b	34.16 bc
1.0% chitosan + 15 mM vanillin	9.02 c	11.66 e
1.5% chitosan + 15 mM vanillin	6.94 c	9.16 e
Storage day	-	-
0	0.00 e	0.00 e
3	0.00 e	0.00 e
6	14.06 d	16.87 d
9	31.16 c	36.25 c
12	54.68 b	51.25 b
15	75.52 a	71.25 a
Interaction Treatment × Storage day	**	**

[z] Mean values in a column followed by different letters indicate significant difference according to LSD at $p < 0.05$.
** Highly significant at $p \leq 0.05$ (n = 24).

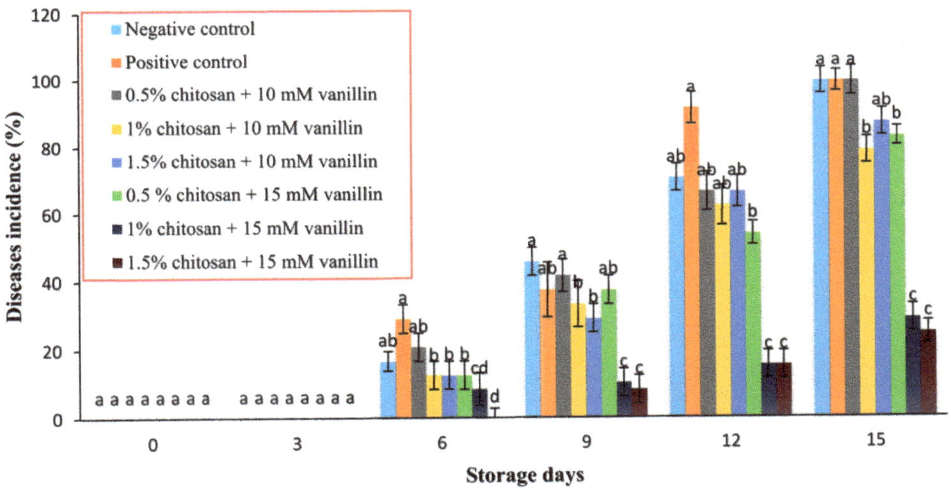

Figure 3. Effects of coating treatment on disease incidence of *Fusarium oxysporum* in tomato fruit stored for 15 days at 26 ± 2 °C and 60 ± 5% relative humidity. Mean values in a column followed by different letters for each storage day differed significantly by LSD at $p \leq 0.05$. Vertical bars indicate standard error of means. Prior to analysis, the data were square root transformed, while non-transformed means are shown (n = 24).

3.2. Total Phenolic Content

There were significant interaction effects of treatment and storage day on the total phenolic content (TPC) of tomato fruit during storage (Table 4).

Figure 5 shows that coating treatment and storage day affected the total phenolic content of tomato fruit. By storage day 3, the fruit treated with 1% chitosan + 15 mM vanillin and 1.5% chitosan + 15 mM vanillin continued to have lower TPC than other treatments. However, at storage day 6, there was no significant difference among treatments. By storage day 9, fruit treated with 1% chitosan + 15 mM vanillin and 1.5% chitosan + 15 mM vanillin obviously had lower TPC than positive and negative control fruit and those treated with 0.5% chitosan + 10 mM vanillin, 1% chitosan + 10 mM vanillin, 1.5% chitosan + 10 mM vanillin and 0.5% chitosan + 15 mM vanillin. This trend continued until the end of storage day 15.

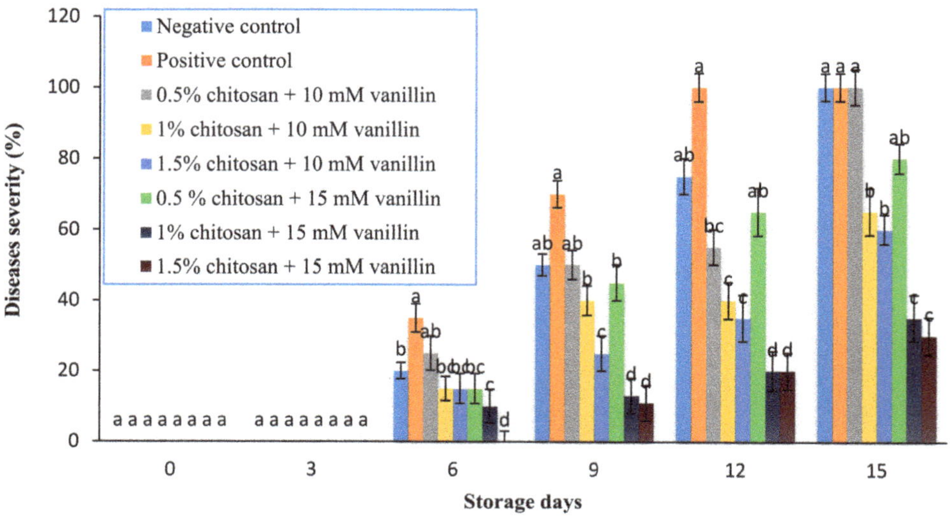

Figure 4. Effects of coating treatment on disease severity of *Fusarium oxysporum* in tomato fruit stored for 15 days at 26 ± 2 °C and 60 ± 5% relative humidity. Mean values in a column followed by different letters for each storage day differed significantly by LSD at $p \leq 0.05$. Vertical bars indicate standard error of means. Prior to analysis, the data were square root transformed, while non-transformed means are shown (n = 24).

Table 3. Pearson's correlation coefficients for disease incidence and severity of *Fusarium oxysporum* inoculated tomato fruit stored at 26 ± 2 °C and 60 ± 5% relative humidity for 15 days.

	Disease Incidence	Disease Severity
Disease incidence	-	-
Disease severity	0.94 **	-

** Significant correlation at $p \leq 0.05$ and $p \leq 0.01$ (n = 24).

Table 4. Main and interaction effects of different coating treatments and storage days on antioxidant capacity of *Fusarium oxysporum* inoculated tomato fruit stored at 26 ± 2 °C 60 ± 5% relative humidity for 15 days.

Factor	Total Phenolic Content (mg GAE/100 g FW)	FRAP (mM TE/g FW)	ABTS (% Inhibition)
Treatment	-	-	-
Negative control	48.08 a [z]	1615.98 a	37.99 a
Positive control	51.61 a	1562.32 a	40.22 a
0.5% chitosan + 10 mM vanillin	48.31 a	1639.26 a	36.32 ab
1% chitosan + 10 mM vanillin	47.98 a	1655.78 a	36.81 ab
1.5% chitosan + 10 mM vanillin	43.86 b	1418.87 b	34.23 b
0.5% chitosan + 15 mM vanillin	46.38 ab	1567.28 a	36.69 ab

Table 4. *Cont.*

Factor	Total Phenolic Content (mg GAE/100 g FW)	FRAP (mM TE/g FW)	ABTS (% Inhibition)
1.0% chitosan + 15 mM vanillin	36.64 c	1315.24 c	29.82 c
1.5% chitosan + 15 mM vanillin	34.88 c	1287.16 c	28.60 c
Storage day	-	-	-
0	33.65 c	1703.27 b	31.89 b
3	34.11 bc	1625.36 b	29.14 c
6	42.11 b	1668.51 b	33.15 ab
9	44.51 a	1256.06 c	34.76 ab
12	43.57 a	1002.46 d	37.48 a
15	47.88 a	1994.8 a	39.80 a
Interaction Treatment × Storage day	**	**	*

[z] Mean values in column followed by different letters indicate significant difference according to LSD at $p < 0.05$, * significant at $p \leq 0.05$, ** significant at $p \leq 0.05$, [ns] non-significant at $p \leq 0.05$ (n = 24).

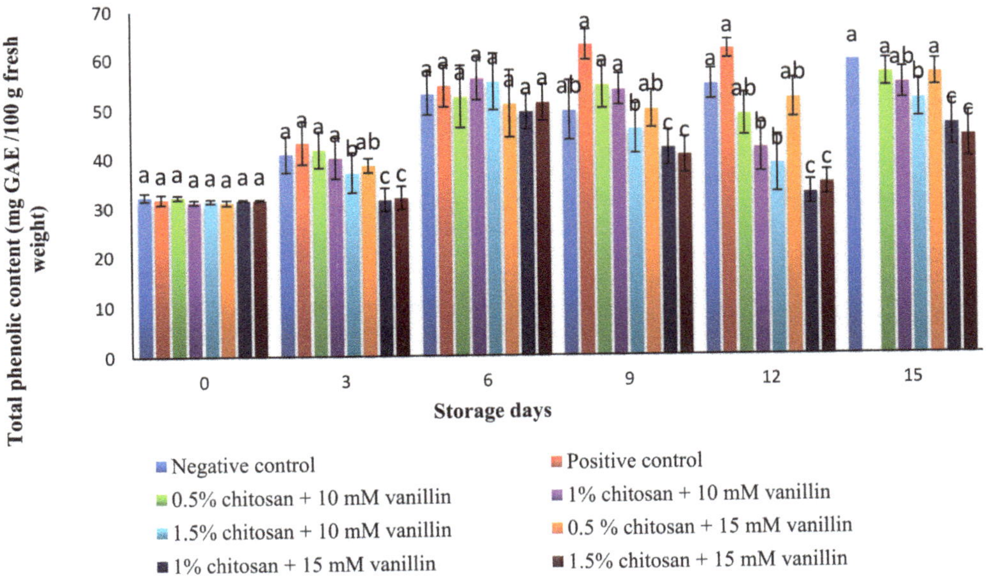

Figure 5. Effects of coating treatment on total phenolic content in *Fusarium oxysporum* inoculated tomato fruit stored for 15 days at 26 ± 2 °C and 60 ± 5% relative humidity. Mean values in a column followed by different letters for each storage day differed significantly by LSD at $p \leq 0.05$. Vertical bars indicate standard error of means (n = 24).

3.3. Antioxidant Capacity (FRAP and ABTS)

Table 4 indicates that there were highly significant interaction effects between coating treatments and storage duration of tomato fruit on the FRAP. At day 3, Figure 6 shows fruit coated with 1.5% chitosan + 10 mM vanillin, 1% chitosan + 15 mM vanillin and 1.5% chitosan + 15 mM vanillin had lower FRAP than positive and negative control fruit and also those coated with 0.5% chitosan + 10 mM vanillin, 1% chitosan + 10 mM vanillin and 0.5% chitosan + 15 mM vanillin. By day 6, fruit with 0.5% chitosan + 15 mM vanillin and 1% chitosan + 10 mM vanillin had greater FRAP than other fruit. Nevertheless, at day 9, there was no significant difference among treatment on fruit FRAP. This trend continued until day 12. By day 15 fruit coated with 1% chitosan + 15 mM vanillin and 1.5% chitosan + 15 mM vanillin had lower FRAP than negative control fruit and those coated with 0.5% chitosan + 10 mM vanillin, 1% chitosan + 10 mM vanillin, 1.5% chitosan + 10 mM vanillin and 0.5% chitosan + 15 mM vanillin.

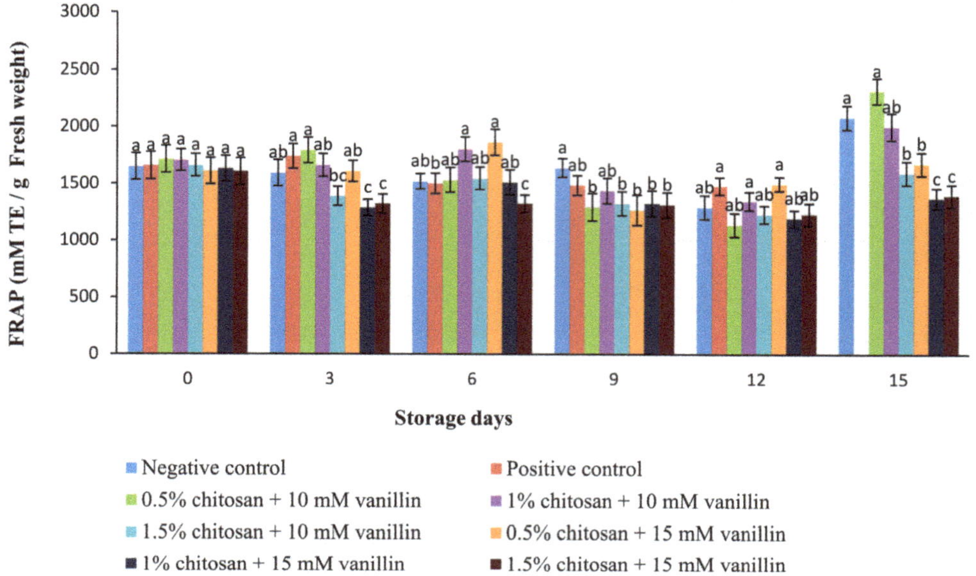

Figure 6. Effects of coating treatment on FRAP in *Fusarium oxysporum* inoculated tomato fruit stored for 15 days at 26 ± 2 °C and 60 ± 5% relative humidity. Mean values in a column followed by different letters in each storage day differed significantly by LSD at $p \leq 0.05$. Vertical bars indicate standard error of means ($n = 24$).

The interaction was significant between treatments and storage days in tomato fruit ABTS, 2, 2-azino-bis (3-ethylbenzthiazoline-6-sulfonic acid) (Table 4). According to the results as shown in Figure 7, there was no significant difference in ABTS among treatment at day 0. By storage day 3, fruit treated with 1.5% chitosan + 10 mM vanillin and 1.5% chitosan + 15 mM vanillin showed lower ABTS radical scavenging capacity than other treatments. At day 6, fruit coated with 1% chitosan + 15 mM vanillin and 1.5% chitosan + 15 mM vanillin had lower ABTS radical scavenging capacity than the control fruit and those treated with 0.5% chitosan + 10 mM vanillin, 1% chitosan + 10 mM vanillin, 1.5% chitosan + 10 mM vanillin and 0.5% chitosan + 15 mM vanillin. A similar trend was found in fruit stored for 9, 12 and 15 days.

There was a significant positive correlation between antioxidant (TPC) and antioxidant capacity (ABTS and FRAP) in tomato fruit treated with chitosan and vanillin during entire storage. From Pearson's correlation analysis, there was a significant positive correlation between TPC and ABTS ($r = 0.53$) and FRAP ($r = 0.76$). There was also a significant positive correlation between FRAP and ABTS ($r = 0.62$) (Table 5).

3.4. Defense-Related Enzyme (PAL, PPO and POD) Activity

In the present study, there were significant interaction effects between coating treatments and storage days in PAL, PPO and POD activities of tomato fruit (Table 6).

Figure 8 shows that there were no significant changes in PAL enzyme activity among treatments at day 0. At storage day 3, the activity of the enzyme dropped slightly in all fruit. However, by day 3, fruit treated with 1% chitosan + 15 mM vanillin and 1.5% chitosan + 15 mM vanillin showed lower PAL activity than positive and negative control fruit and also those coated with 0.5% chitosan + 10 mM vanillin, 1% chitosan + 10 mM vanillin, 1.5% chitosan + 10 mM vanillin and 0.5% chitosan + 15 mM vanillin. This trend continued until the end of storage day 15.

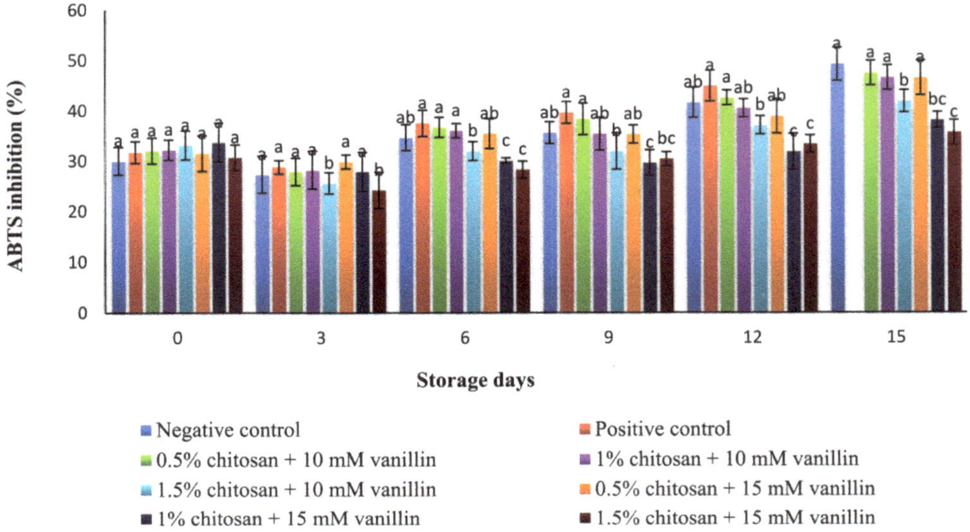

Figure 7. Effects of coating treatment on ABTS in *Fusarium oxysporum* inoculated tomato fruit stored for 15 days at 26 ± 2 °C and 60 ± 5% relative humidity. Mean values in a column followed by different letters for each storage day differed significantly by LSD at $p \leq 0.05$. Vertical bars indicate standard error of means (n = 24).

Table 5. Pearson's correlation coefficients for TPC, DPPH, ABTS and FRAP of *Fusarium oxysporum* inoculated tomato fruit stored at 26 ± 2 °C and 60 ± 5% relative humidity during 15 days of storage.

	TPC	ABTS	FRAP
TPC	-	-	-
ABTS	0.53 **	-	-
FRAP	0.76 **	0.62 **	-

TPC = Total phenolic content, DPPH = 2,2-diphenyl-1-picrylhydrazyl, ABTS = 2,2′-azino-bis(3-ethylbenzothiazoline-6-sulphonic acid) and FRAP = ferric reducing antioxidant power. ** Significant correlation at $p \leq 0.05$ and $p \leq 0.01$ (n = 24).

Table 6. Main and interaction effects of different coating treatments and storage days on defense-related enzyme activity of *Fusarium oxysporum* inoculated tomato fruit stored at 26 ± 2 °C and 60 ± 5% relative humidity for 15 days.

Factor	PAL Specific Activity (U/mg protein)	PPO Specific Activity (U/mg protein)	POD Specific Activity (U/mg protein)
Treatment	-	-	-
Negative control	0.48 ab [z]	0.64 b	1.04 ab
Positive control	0.53 a	0.78 a	1.10 a
0.5% chitosan + 10 mM vanillin	0.46 ab	0.66 ab	1.06 ab
1% chitosan + 10 mM vanillin	0.43 b	0.58 c	1.01 ab
1.5% chitosan + 10 mM vanillin	0.39 c	0.64 b	1.04 ab
0.5% chitosan + 15 mM vanillin	0.46 ab	0.58 c	0.96 bc
1.0% chitosan + 15 mM vanillin	0.35 d	0.52 d	0.86 c
1.5% chitosan + 15 mM vanillin	0.34 d	0.54 d	0.91 c
Storage day	-	-	-
0	0.45 ab	0.49 d	0.74 d
3	0.46 ab	0.55 c	1.05 b
6	0.42 b	0.52 cd	0.87 c
9	0.45 ab	0.73 b	0.93 bc
12	0.48 a	0.72 b	1.24 a

Table 6. *Cont.*

15	0.58 a	0.84 a	1.57 a
Interaction Treatment × Storage day	**	**	**

[z] Mean values in a column followed by different letters indicate significant difference according to LSD at $p < 0.05$.
** Highly significant at $p \leq 0.05$ ($n = 24$).

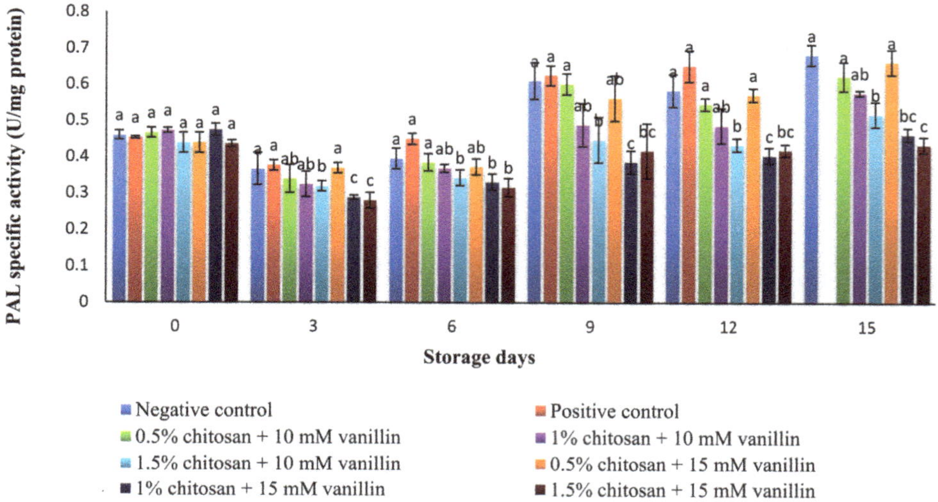

Figure 8. Effects of coating treatment on PAL specific activity in *Fusarium oxysporum* inoculated tomato fruit stored for 15 days at 26 ± 2 °C and 60 ± 5% relative humidity. Mean values in a column followed by different letters for each storage day differed significantly by LSD at $p \leq 0.05$. Vertical bars indicate standard error of means ($n = 24$).

Figure 9 shows that the PPO enzyme activity of tomato fruit at storage day 6 that was treated with 1% chitosan + 15 mM vanillin and 1.5% chitosan + 15 mM vanillin had lower PPO activity than positive and negative control fruit and also those coated with 0.5% chitosan + 10 mM vanillin, 1% chitosan + 10 mM vanillin, 1.5% chitosan + 10 mM vanillin and 0.5% chitosan + 15 mM vanillin. This trend continued until the end of storage day 15.

Figure 10 exhibits POD enzymes activity of tomato fruit increased slightly as storage day advanced to 3. Nevertheless, fruit treated with 1% chitosan + 15 mM vanillin and 1.5% chitosan + 15 mM vanillin shows lower POD activity than other treatments. By storage day 6, fruit coated with 1.5% chitosan + 10 mM vanillin, 1% chitosan + 15 mM vanillin and 1.5% chitosan + 15 mM vanillin shows lowest POD activity as compared to others treatment. This trend continued for the rest of storage day 15, however, the POD enzymes activity increased slightly in all treated fruit after 9 day of storage.

There was a significant correlation among defense-related enzymes. Pearson's correlation analysis shows that there was a highly significant positive correlation between PAL and PPO (r = 0.82), intermediate correlation of PAL and POD (r = 0.74) and intermediate correlation between POD and PPO (r = 0.67) (Table 7).

Table 7. Pearson's correlation coefficients for disease incidence and severity of *Fusarium oxysporum* inoculated tomato fruit stored at 26 ± 2 °C and 60 ± 5% relative humidity for 15 days.

	PAL	PPO	POD
PAL	-	-	-
PPO	0.82 **	-	-
POD	0.74 **	0.67 **	-

PAL = Phenylalanine ammonia-lyase, POD = Peroxidase and PPO = Polyphenoloxidase. ** Significant correlation at $p \leq 0.05$ and $p \leq 0.01$ ($n = 24$).

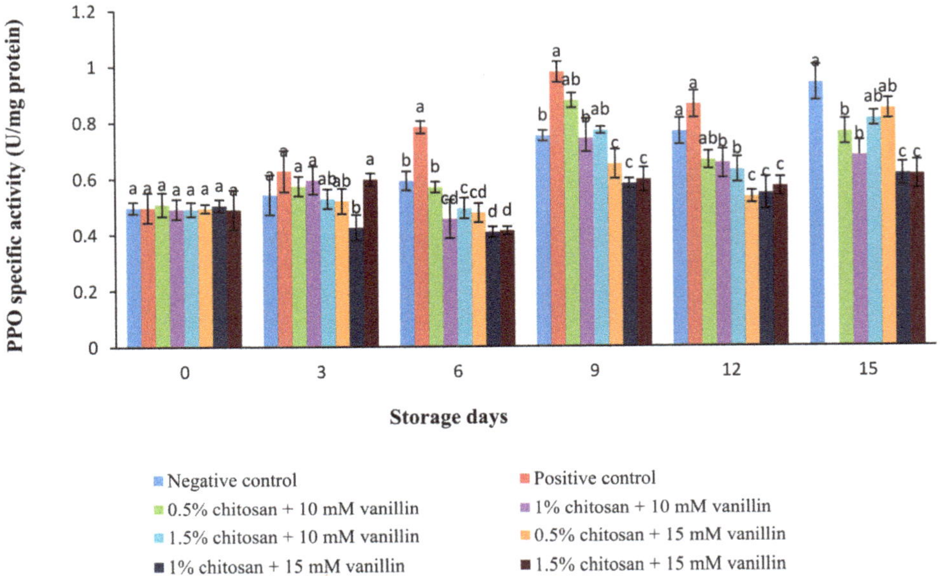

Figure 9. Effects of coating treatment on PPO specific activity in *Fusarium oxysporum* inoculated tomato fruit stored for 15 days at 26 ± 2 °C and 60 ± 5% relative humidity. Mean values in a column followed by different letters for each storage day differed significantly by LSD at $p \leq 0.05$. Vertical bars indicate standard error of means (n = 24).

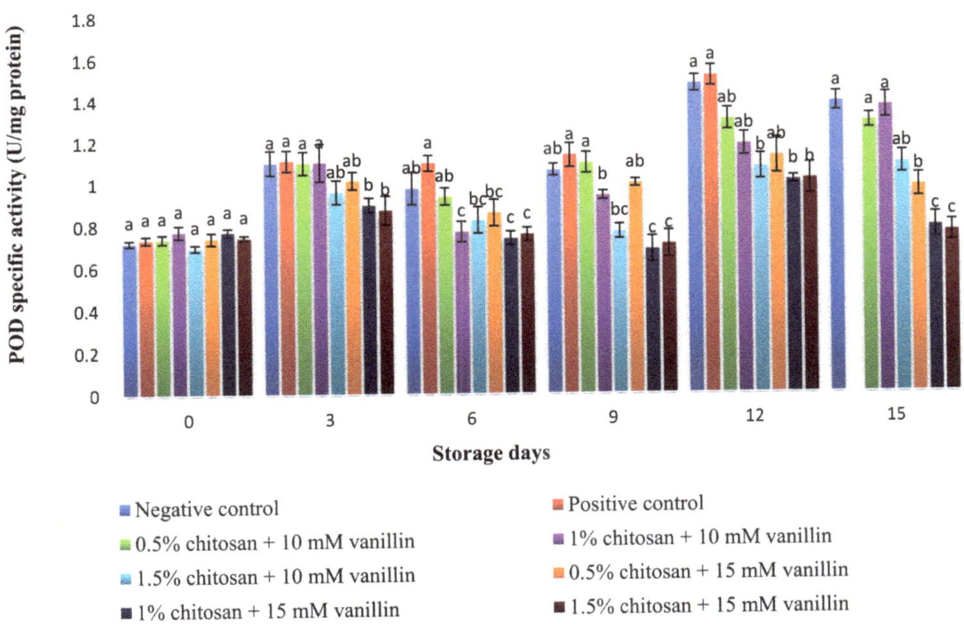

Figure 10. Effects of coating treatment on POD specific activity in *Fusarium oxysporum* inoculated tomato fruit stored for 15 days at 26 ± 2 °C and 60 ± 5% relative humidity. Mean values in a column followed by different letters for each storage day differed significantly by LSD at $p \leq 0.05$. Vertical bars indicate standard error of means (n = 24).

4. Discussion

4.1. Disease Incidence and Severity

The present study revealed that disease incidence and severity increased as the storage period advanced, while the coating significantly affected the percentage of disease incidence and severity during the storage period. As the storage days progressed, the control fruit and those treated with a low concentration of chitosan and vanillin showed more severe infection. In contrast, fruit coated with a high concentration of chitosan and vanillin had an inhibited progression of the disease in tomato fruit, as seen in Figures 3 and 4. Most probably, the chitosan coating formed a semi-permeable film around the fruit; this thin film could inhibit the growth of pathogens by disturbing the cell membrane of the pathogen that causes intracellular leakage and finally cell death. In addition, the chitosan coating can enhance the epidermal structure of fruit and limit the spread of the pathogens. Abebe and Mohammed [16,35] expressed that the coating could assist the cell wall in retaining its integrity against fungal attack and help in delaying pathogenic infection. This result was in agreement with the findings of Chen [36], where disease incidence and severity were lower in 1.5% chitosan-coated navel oranges than 0.5% chitosan when stored for 120 days at 5 ± 0.5 °C and 85–90% RH. Sikder [37] found that disease incidence and severity of bananas coated with 1% chitosan were lower as compared to 0.5% chitosan during 12 days storage at 28 ± 2 °C. In papayas, 1% chitosan significantly reduced anthracnose disease incidence and severity by 80% as compared to 0.05% chitosan when stored for 14 days at 13.5 °C and 96% RH [38]. In the present study, the barrier formed by the higher concentration of coating could have inhibited the growth of pathogens and slowed down the ripening and senescence process of tomato fruit, and therefore the disease incidence and severity were lesser in this fruit.

From the Pearson's correlation analysis, there was strong significant positive correlation between disease incidence and severity ($r = 0.94$) (Table 3). This was in agreement with Rashid [39], who found high correlation between disease incidence and severity ($r = 0.91$) in papaya fruit during 15 days of storage. In line with this study, Hossain [40] also found highly significant positive correlation between disease incidence and severity ($r = 0.89$) in banana fruit that was coated with 0.5, 0.75 and 1% chitosan and stored at 26 ± 2 °C and 85 ± 5% RH for 4 days. It is clear that disease incidence is a main contributor to disease severity.

4.2. Total Phenolic Content

Phenolic compounds, or secondary metabolites, are widely distributed in plants. They are particularly involved in plant defense against ultraviolet radiation and aggression by a pathogen [41]. Phenolic compounds are probably the most important candidates contributing to the antioxidant properties of plants and are associated with the scavenging of free radicals, breaking radical chain reactions, and chelating metals [42]. In this study, coating treatments had a significant effect on total phenolic content over the entire storage period. Their interaction effect was significant between treatment and storage day on the TPC of tomato fruit (Figure 5). In general, fruit treated with higher concentrations of chitosan and vanillin had lower TPC than those coated with low concentrations of chitosan and vanillin. However, Figure 3 illustrates that fruit treated with 1.5% chitosan + 15 mM vanillin had 25.5% lower TPC as compared to fruit coated with 0.5% chitosan + 1 mM vanillin at the end of storage day 15.

This lower TPC might be because the control fruit and those coated with low concentrations of chitosan ripened faster, and the phenolic compounds might have reacted with other compounds. It also appeared that the rise in phenolic levels may be due to biotic stresses, degradation of cells, and senescence [43]. In agreement with this study Munhuweyi [10] reported that pomegranate fruit coated with 1.5% chitosan had 36% lower total phenolic content when stored for 14 days at 4 °C as compared to fruit coated with 0.5% chitosan. In line with this study, previous researchers reported that a 1.5% chitosan coating caused a greater reduction in TPC as compared to fruit coated with 0.5% chitosan, as found in sweet cherries [44], cut pineapple [45] and blueberries [46]. In the current study, the film created by the higher concentrations of coating slowed down the ripening and senescence process and suppressed abiotic stress in the fruit, modifying its metabolism and resulting in lower TPC.

4.3. Antioxidant Activity and Capacity

A number of assays have been introduced for the measurement of the total antioxidant activity of fruit [47]. In recent years, a wide range of spectrophotometric assays has been adopted to measure the antioxidant capacity of foods. The most popular are 2,2′-azino-bis-3-ethylbenzthiazoline-6-sulphonic acid (ABTS), 2, and ferric reducing antioxidant power (FRAP) [48]. Most of the assays employ the same principle: a synthetic colored radical or redox-active compound is generated, and the ability of a biological sample to scavenge the radical or to reduce the redox-active compound is monitored by a spectrophotometer.

Figure 6 shows that fruit coated with 1.5% chitosan + 15 mM vanillin had 30.45% lower FRAP than fruit coated with 0.5% chitosan + 10 mM vanillin at the end of storage day 15. The lower FRAP in tomato fruit coated with a higher concentration of chitosan and vanillin might be due to the formation of a protective barrier on the surface of fresh fruit that inhibits and reduces fruit antioxidant activity. Many researchers have reported that FRAP decreases during storage when fruit is coated using higher concentrations of chitosan, as occurred in pomegranates [49], strawberries [50] and tomatoes [51].

Figure 7 shows the ABTS of tomato fruit decreased as the concentration of chitosan and vanillin increased, while advancement of storage day increased its ABTS. Tomato fruit coated with 1.5% chitosan + 15 mM vanillin showed 19.66% lower ABTS than fruit coated with 0.5% chitosan + 10 mM vanillin. In line with this study, Martínez found that strawberry fruit coated with 1.5% chitosan had lower ABTS than fruit coated [52] with 0.5% chitosan during 15 days of storage. Most probably, the barrier formed by the higher concentration of coating delayed the senescence process and reduced decay in tomato fruit, and thus ABTS was lower in this fruit.

From Pearson's correlation analysis, there was a significant positive correlation between TPC and ABTS (r = 0.53) and FRAP (r = 0.76). There was also a significant positive correlation between FRAP and ABTS (r = 0.62) (Table 5). This was in agreement with Sushant et al. (2019) who found highly positive correlation between TPC and DPPH (r = 0.75) in Cassia tora plant. Similar result was also reported by Floegel et al. (2011) that there was highly significant in apple fruit that the correlation between TPC and DPPH (r = 0.89), highly significant correlation TPC and ABTS (r = 0.94) and strong significant correlation of TPC and FRAP (r = 0.70). In line with this study, Fu et al. (2010) also found highly significant positive correlation between TPC and antioxidant capacity (FRAP) (r = 0.79) in *Ficus benjamina*. The finding of this study indicated that TPC is the major contributor for tomato fruit antioxidant capacity.

4.4. Effects of Coating on the Activity of Defense-Related Enzymes (PAL, PPO and POD)

PAL, PPO and POD are among the most important enzymes having defensive responses in plants against insects and pathogens [53]. PAL is a key enzyme in the metabolism of phenols that protect plants against stress conditions [54]. There was a significant interaction effect between treatment and storage day on tomato fruit defensive enzyme PAL activity (Figure 8). At the end of storage day 15, PAL activity of fruit coated with 1.5% chitosan + 15 mM vanillin was 44.18% lower than those coated with 0.5% chitosan + 10 mM vanillin. A study by Zhan and Zhu [55] found that the PAL activity of water caltrop fresh fruit (*Trapa natans* L.) coated with 1% and 2% chitosan was lower than those coated with 0.5% chitosan during 15 days of storage at 4 ± 1 °C and 80%–85% RH. Previous researchers also reported that jujube fruit (*Ziziphus jujuba* Mill.) [56] and tomato fruit [57] with 1.5% chitosan coating had lower PAL activity than those coated with 0.5%. In the present study, the layer created by the higher concentration of coating most probably reduced ethylene production rate and thus slowed down the ripening and senescence process of tomato fruit, leading to low PAL activity.

PPO is a key defense enzyme against pathogen reaction through the oxidation of polyphenols into quinines, which have antimicrobial activity and also strengthen the resistance of plant cells during microbial attack [27,58]. Figure 9 shows that, at storage day 15, the PPO activity of tomato fruit coated with 1.5% chitosan + 15 mM vanillin was 21.4% lower than the PPO activity of fruit coated with 0.5% chitosan + 10 mM vanillin. In agreement with this study, Minh [59] found the PPO activity of fresh mushrooms coated in 1.5% chitosan was lower than those coated with 0.5% chitosan. A study by Ghasemnezhad [49] demonstrated that PPO activity in pomegranate fruit coated with 1% chitosan was lower than those coated with 0.5% chitosan. A similar finding was also reported in litchi fruit [60] and tomato fruit [61] during storage. The reduction of PPO activity in high concentration chitosan coated tomato fruit might be due to low respiration and ethylene production rates, reducing disease attack and slowing ripening and senescence processes.

POD is one of the enzymes expressed in different stimuli, including pathogenic challenges, and has important roles in pathogenesis, oxidative burst, and resistance to infection [62]. As the storage day of tomato fruit advanced, fruit POD activity increased; in contrast, as the concentration of chitosan and vanillin coating increased, the POD activity decreased (Figure 10). However, fruit coated with 1.5% chitosan + 15 mM vanillin had 40.9% lower POD activity than fruit coated with 0.5% chitosan + 10 mM vanillin at the end of storage day 15. In line with this study, Ismail [63] found that fresh green beans coated with 1.5% chitosan had lower POD than those coated with 0.5% chitosan stored at 4 °C and 85%–90% RH for 28 days. In agreement with this study, previous researchers reported that 1.5% chitosan had lower POD in fruit than those coated with 0.5% chitosan, as found in tomato fruit [5], mushrooms [60] and strawberries [50]. In the current study, the film formed by the higher concentration of coating reduced disease attack and cell structure damaged by the pathogen and also slowed down respiration rate, ripening and senescence processes of tomato fruit; thus POD activity was lower in this fruit.

Pearson's correlation analysis shows that there was a highly significant positive correlation between PAL and PPO ($r = 0.82$), intermediate correlation of PAL and POD ($r = 0.74$) and intermediate correlation between POD and PPO ($r = 0.67$) (Table 7). The result was in agreement with Adiletta [64], who found higher correlation between PPO and POD ($r = 0.79$) in loquat fruit coated with 1% chitosan and stored at 7 °C for 21 days. In line with this study, Pasquariello [65] also found a highly positive correlation between PPO and POD ($r = 0.87$) and PPO and PAL ($r = 0.71$) in strawberry fruit coated with 1% chitosan stored at 2 °C and 95% RH for 14 days. This result indicated that defense-related enzymes such as PAL, PPO and POD are the main contributor to the oxidation of polyphenols into quinines, which strengthen the resistance of the plant cells during microbial attack.

5. Conclusions

The chitosan and vanillin coating could be considered as a commercial application to improve shelf life and maintain tomato fruit quality during storage at a room temperature of 26 ± 2 °C and at 60 ± 5% RH. The present findings show that chitosan and vanillin coating can effectively inhibit postharvest diseases in tomato fruit by controlling the disease incidence and severity as well as by keeping constant the defense-related enzyme activity. Furthermore, chitosan and vanillin consistently maintain the antioxidant activity and capacity. Our results suggest that a chitosan and vanillin coating of 1.5% chitosan + 15 mM vanillin formed a protective layer on fruit surfaces that helped to inhibit disease occurrence, slowing down the ripening and senescence processes in tomato fruit. As a result, tomato fruit effectively stored under 26 ± 2 °C and 60 ± 5% RH for 15 days, even inoculated with *Fusarium oxysporum*.

Author Contributions: Data curation, Z.S.S. and S.F.Y.; formal analysis, Z.S.S. and S.F.Y.; funding acquisition, P.D.; investigation, Z.S.S.; methodology, Z.S.S.; project administration, P.D. and J.J.N.;

resources, Z.S.S. and P.D.; supervision, P.D.; writing—original draft, Z.S.S.; writing—review & editing, P.D., J.J.N. and S.F.Y. All authors have read and agreed to the published version of the manuscript.

Funding: This research received no external funding. The research was supported by supervisor Phebe Ding and the publication fee was funded by UPM Research Management Center.

Institutional Review Board Statement: Not applicable.

Informed Consent Statement: Not applicable.

Data Availability Statement: The data used to support the findings of this study are included within the article.

Conflicts of Interest: The authors declare that they have no known competing financial interests or personal relationships that could have appeared to influence the work reported in this paper.

References

1. Garuba, T.; Mustapha, O.; Oyeyiola, G. Shelf life and proximate composition of tomato (*Solanum lycopersicum* L.) fruits as influenced by storage methods. *Ceylon J. Sci.* **2018**, *47*, 387. [CrossRef]
2. Sun, C.; Jin, L.; Cai, Y.; Huang, Y.; Zheng, X.; Yu, T. l-Glutamate treatment enhances disease resistance of tomato fruit by inducing the expression of glutamate receptors and the accumulation of amino acids. *Food Chem.* **2019**, *293*, 263–270. [CrossRef]
3. Elive, L.L. Postharvest Losses of Tomato (*Lycopersicum esculentum*) and Handling Practices in Wotuto, South West Region Cameron. Ph.D. Thesis, Pan African Institute for Development West Africa Cameroon, Buea, Cameroon, 2018.
4. Emana, B.; Afari-Sefa, V.; Nenguwo, N.; Ayana, A.; Kebede, D.; Mohammed, H. Characterization of pre- and postharvest losses of tomato supply chain in Ethiopia. *Agric. Food Secur.* **2017**, *6*, 87. [CrossRef]
5. Liu, X.; Gao, Y.; Yang, H.; Li, L.; Jiang, Y.; Li, Y.; Zheng, J. Pichia kudriavzevii retards fungal decay by influencing the fungal community succession during cherry tomato fruit storage. *Food Microbiol.* **2020**, *88*, 103404. [CrossRef] [PubMed]
6. Bakar, A.A.; Izzati, M.N.A.; Kalsom, Y. Diversity of Fusarium species associated with postharvest fruit rot disease of tomato. *Sains Malays.* **2013**, *42*, 911–920.
7. Bouzoumita, A.; Metoui, M.; Jemni, M.; Kabaeir, N.; Belhouchette, K.; Ferchichi, A. The Efficacy of Various Bacterial Organisms for Biocontrol of Fusarium Root Rot of Olive in Tunisia. *Pol. J. Environ. Stud.* **2019**, *29*, 11–16. [CrossRef]
8. Medina-Romero, Y.M.; Roque-Flores, G.; Macías-Rubalcava, M.L. Volatile organic compounds from endophytic fungi as innovative postharvest control of Fusarium oxysporum in cherry tomato fruits. *Appl. Microbiol. Biotechnol.* **2017**, *101*, 8209–8222. [CrossRef] [PubMed]
9. Zheng, F.; Zheng, W.; Li, L.; Pan, S.; Liu, M.; Zhang, W.; Liu, H.; Zhu, C. Chitosan Controls Postharvest Decay and Elicits Defense Response in Kiwifruit. *Food Bioprocess Technol.* **2017**, *10*, 1937–1945. [CrossRef]
10. Munhuweyi, K.; Lennox, C.L.; Meitz-Hopkins, J.C.; Caleb, O.J.; Sigge, G.O.; Opara, U.L. Investigating the effects of crab shell chitosan on fungal mycelial growth and postharvest quality attributes of pomegranate whole fruit and arils. *Sci. Hortic.* **2017**, *220*, 78–89. [CrossRef]
11. Liu, J.; Tian, S.; Meng, X.; Xu, Y. Effects of chitosan on control of postharvest diseases and physiological responses of tomato fruit. *Postharvest Biol. Technol.* **2007**, *44*, 300–306. [CrossRef]
12. Kuwar, U.; Sharma, S.; Tadapaneni, V.R.R. A loe Vera Gel and Honey-Based Edible Coatings Combined with Chemical Dip as a Safe Means for Quality Maintenance and Shelf Life Extension of Fresh-Cut Papaya. *J. Food Qual.* **2015**, *38*, 347–358. [CrossRef]
13. Konagaya, K.; Al Riza, D.F.; Nie, S.; Yoneda, M.; Hirata, T.; Takahashi, N.; Kuramoto, M.; Ogawa, Y.; Suzuki, T.; Kondo, N. Monitoring mature tomato (red stage) quality during storage using ultraviolet-induced visible fluorescence image. *Postharvest Biol. Technol.* **2020**, *160*, 111031. [CrossRef]
14. Ali, A.; Maqbool, M.; Ramachandran, S.; Alderson, P.G. Gum arabic as a novel edible coating for enhancing shelf-life and improving postharvest quality of tomato (*Solanum lycopersicum* L.) fruit. *Postharvest Biol. Technol.* **2010**, *58*, 42–47. [CrossRef]
15. Khaliq, G.; Mohamed, M.T.M.; Ali, A.; Ding, P.; Ghazali, H.M. Effect of gum arabic coating combined with calcium chloride on physico-chemical and qualitative properties of mango (*Mangifera indica* L.) fruit during low temperature storage. *Sci. Hortic.* **2015**, *190*, 187–194. [CrossRef]
16. Mohamed, N.T.S.; Ding, P.; Ghazali, H.M.; Kadir, J. Biochemical and cell wall ultrastructural changes in crown tissue of banana (Musa AAA 'Berangan') fruit as mediated by UVC irradiation against crown rot fungal infection. *Postharvest Biol. Technol.* **2017**, *128*, 144–152. [CrossRef]
17. Rastegar, S.; Khankahdani, H.H.; Rahimzadeh, M. Effectiveness of alginate coating on antioxidant enzymes and biochemical changes during storage of mango fruit. *J. Food Biochem.* **2019**, *43*, e12990. [CrossRef]
18. Si, X.; Chen, Q.; Bi, J.; Wu, X.; Yi, J.; Zhou, L.; Li, Z. Comparison of different drying methods on the physical properties, bioactive compounds and antioxidant activity of raspberry powders. *J. Sci. Food Agric.* **2016**, *96*, 2055–2062. [CrossRef]
19. Zainal, B.; Ding, P.; Ismail, I.S.; Saari, N. 1H NMR metabolomics profiling unveils the compositional changes of hydro-cooled rockmelon (*Cucumis melo* L. reticulatus cv glamour) during storage related to in vitro antioxidant activity. *Sci. Hortic.* **2019**, *246*, 618–633. [CrossRef]

20. Dave, P.N. Evaluation of Antioxidant Activities by use of Various Extracts from Abutilon Pannosum and Grewia Tenax in the Kachchh Region. *MOJ Food Process. Technol.* **2017**, *5*, 1–13. [CrossRef]
21. Pinheiro, J.C.; Alegria, C.S.M.; Abreu, M.M.M.N.; Gonçalves, E.M.; Silva, C.L.M. Evaluation of Alternative Preservation Treatments (Water Heat Treatment, Ultrasounds, Thermosonication and UV-C Radiation) to Improve Safety and Quality of Whole Tomato. *Food Bioprocess Technol.* **2016**, *9*, 924–935. [CrossRef]
22. Briones-Labarca, V.; Giovagnoli-Vicuña, C.; Cañas-Sarazúa, R. Optimization of extraction yield, flavonoids and lycopene from tomato pulp by high hydrostatic pressure-assisted extraction. *Food Chem.* **2019**, *278*, 751–759. [CrossRef]
23. Thaipong, K.; Boonprakob, U.; Crosby, K.; Cisneros-Zevallos, L.; Byrne, D.H. Comparison of ABTS, DPPH, FRAP, and ORAC assays for estimating antioxidant activity from guava fruit extracts. *J. Food Compos. Anal.* **2006**, *19*, 669–675. [CrossRef]
24. Jumnongpon, R.; Chaiseri, S.; Hongsprabhas, P.; Healy, J.; Meade, S.; Gerrard, J. Cocoa protein crosslinking using Maillard chemistry. *Food Chem.* **2012**, *134*, 375–380. [CrossRef]
25. Raseetha, S.; Heenan, S.; Oey, I.; Burritt, D.; Hamid, N. A new strategy to assess the quality of broccoli (*Brassica oleracea* L. italica) based on enzymatic changes and volatile mass ion profile using Proton Transfer Reaction Mass Spectrometry (PTR-MS). *Innov. Food Sci. Emerg. Technol.* **2011**, *12*, 197–205. [CrossRef]
26. Bonjoch, N.P.; Tamayo, P.R. Protein content quantification by Bradford method. In *Handbook of Plant Ecophysiology Techniques*; Springer: Berlin/Heidelberg, Germany, 2001; pp. 283–295. [CrossRef]
27. Wang, F.; Deng, J.; Jiao, J.; Lu, Y.; Yang, L.; Shi, Z. The combined effects of Carboxymethyl chitosan and Cryptococcus laurentii treatment on postharvest blue mold caused by Penicillium italicum in grapefruit fruit. *Sci. Hortic.* **2019**, *253*, 35–41. [CrossRef]
28. Han, C.; Zuo, J.; Wang, Q.; Xu, L.; Zhai, B.; Wang, Z.; Dong, H.; Gao, L. Effects of chitosan coating on postharvest quality and shelf life of sponge gourd (*Luffa cylindrica*) during storage. *Sci. Hortic.* **2014**, *166*, 1–8. [CrossRef]
29. Habibi, F.; Ramezanian, A.; Rahemi, M.; Eshghi, S.; Guillén, F.; Serrano, M.; Valero, D. Postharvest treatments with γ-aminobutyric acid, methyl jasmonate, or methyl salicylate enhance chilling tolerance of blood orange fruit at prolonged cold storage. *J. Sci. Food Agric.* **2019**, *99*, 6408–6417. [CrossRef]
30. Zhang, J.; Liu, J.; Xie, J.; Deng, L.; Yao, S.; Zeng, K. Biocontrol efficacy of Pichia membranaefaciens and Kloeckera apiculata against *Monilinia fructicola* and their ability to induce phenylpropanoid pathway in plum fruit. *Biol. Control* **2019**, *129*, 83–91. [CrossRef]
31. Kokkinakis, D.M.; Brooks, J.L. Tomato Peroxidase. *Plant Physiol.* **1979**, *63*, 93–99. [CrossRef]
32. Ogola, H.J.O.; Kamiike, T.; Hashimoto, N.; Ashida, H.; Ishikawa, T.; Shibata, H.; Sawa, Y. Molecular Characterization of a Novel Peroxidase from the *Cyanobacterium anabaena* sp. Strain PCC 7120. *Appl. Environ. Microbiol.* **2009**, *75*, 7509–7518. [CrossRef]
33. Kumari, Y.A.I.; Vengadaramana, A. Stimulation of Defense Enzymes in Tomato (*Solanum lycopersicum* L.) and Chilli (*Capsicum annuum* L.) in Response to Exogenous Application of Different Chemical Elicitors. *Univ. J. Plant Sci.* **2017**, *5*, 10–15. [CrossRef]
34. Kumar, R.; Sharma, P.K.; Mishra, P.S. A review on the vanillin derivatives showing various biological activities. *Int. J. PharmTech Res.* **2012**, *4*, 266–279.
35. Abebe, Z.; Tola, Y.B.; Mohammed, A. Effects of edible coating materials and stages of maturity at harvest on storage life and quality of tomato (*Lycopersicon esculentum* Mill.) fruits. *Afr. J. Agric. Res.* **2017**, *12*, 550–565. [CrossRef]
36. Chen, C.; Cai, N.; Chen, J.; Peng, X.; Wan, C. Chitosan-Based Coating Enriched with Hairy Fig (*Ficus hirta* Vahl.) Fruit Extract for "Newhall" Navel Orange Preservation. *Coatings* **2018**, *8*, 445. [CrossRef]
37. Sikder, M.B.H.; Islam, M.M. Effect of Shrimp Chitosan Coating on Physico-chemical Properties and Shelf Life Extension of Banana. *Int. J. Eng. Technol. Sci.* **2019**, *6*, 41–54. [CrossRef]
38. Hewajulige, I.; Sivakumar, D.; Sultanbawa, Y.; Wijeratnam, R.W.; Wijesundera, R.L.C. Effect of chitosan coating on the control of anthracnose and overall quality retention of papaya (*Carica papaya* L.) during storage. *Int. Sympos. Papaya* **2005**, *740*, 245–250. [CrossRef]
39. Rashid, M.; Borman, B.; Islam, M.; Shirin, A. Application of alternative treatments to control postharvest fungal infection and shelf life extension of papaya in different maturity stages. *Progress. Agric.* **2019**, *30*, 298–304. [CrossRef]
40. Hossain, M.S.; Iqbal, A. Effect of shrimp chitosan coating on postharvest quality of banana (*Musa sapientum* L.) fruits. *Int. Food Res. J.* **2016**, *23*, 277–288.
41. Upadhyaya, S.; Khatiwora, E.; Bora, D.K. Estimation of total Phenol and flavonoid contents of Citrus limon L Burmf leaves from North Eastern region of India. *J. Drug Deliv. Ther.* **2019**, *9*, 40–42. [CrossRef]
42. Gonçalves, S.; Moreira, E.; Andrade, P.B.; Valentão, P.; Romano, A. Effect of in vitro gastrointestinal digestion on the total phenolic contents and antioxidant activity of wild Mediterranean edible plant extracts. *Eur. Food Res. Technol.* **2019**, *245*, 753–762. [CrossRef]
43. Bal, E. Influence of Chitosan-Based Coatings with UV Irradiation on Quality of Strawberry Fruit During Cold Storage. *Turk. J. Agric. Food Sci. Technol.* **2019**, *7*, 275–281. [CrossRef]
44. Petriccione, M.; Mastrobuoni, F.; Pasquariello, M.S.; Zampella, L.; Nobis, E.; Capriolo, G.; Scortichini, M. Effect of Chitosan Coating on the Postharvest Quality and Antioxidant Enzyme System Response of Strawberry Fruit during Cold Storage. *Foods* **2015**, *4*, 501–523. [CrossRef]
45. Jing, Y.; Huang, J.; Yu, X. Maintenance of the antioxidant capacity of fresh-cut pineapple by procyanidin-grafted chitosan. *Postharvest Biol. Technol.* **2019**, *154*, 79–86. [CrossRef]
46. Yang, G.; Yue, J.; Gong, X.; Qian, B.; Wang, H.; Deng, Y.; Zhao, Y. Blueberry leaf extracts incorporated chitosan coatings for preserving postharvest quality of fresh blueberries. *Postharvest Biol. Technol.* **2014**, *92*, 46–53. [CrossRef]

47. Re, R.; Pellegrini, N.; Proteggente, A.; Pannala, A.; Yang, M.; Rice-Evans, C. Antioxidant activity applying an improved ABTS radical cation decolorization assay. *Free Radic. Biol. Med.* **1999**, *26*, 1231–1237. [CrossRef]
48. Floegel, A.; Kim, D.-O.; Chung, S.-J.; Koo, S.I.; Chun, O.K. Comparison of ABTS/DPPH assays to measure antioxidant capacity in popular antioxidant-rich US foods. *J. Food Compos. Anal.* **2011**, *24*, 1043–1048. [CrossRef]
49. Ghasemnezhad, M.; Zareh, S.; Rassa, M.; Sajedi, R.H. Effect of chitosan coating on maintenance of aril quality, microbial population and PPO activity of pomegranate (*Punica granatum* L. cv. Tarom) at cold storage temperature. *J. Sci. Food Agric.* **2013**, *93*, 368–374. [CrossRef] [PubMed]
50. Wang, S.Y.; Gao, H. Effect of chitosan-based edible coating on antioxidants, antioxidant enzyme system, and postharvest fruit quality of strawberries (Fragaria x aranassa Duch.). *LWT* **2013**, *52*, 71–79. [CrossRef]
51. Ali, A.; Maqbool, M.; Alderson, P.G.; Zahid, N. Effect of gum arabic as an edible coating on antioxidant capacity of tomato (*Solanum lycopersicum* L.) fruit during storage. *Postharvest Biol. Technol.* **2013**, *76*, 119–124. [CrossRef]
52. Martínez, K.; Ortiz, M.; Albis, A.; Castañeda, C.G.G.; Valencia, M.E.; Tovar, C.D.G. The Effect of Edible Chitosan Coatings Incorporated with Thymus capitatus Essential Oil on the Shelf-Life of Strawberry (Fragaria x ananassa) during Cold Storage. *Biomolecules* **2018**, *8*, 155. [CrossRef] [PubMed]
53. Han, Y.; Wang, Y.; Bi, J.-L.; Yang, X.-Q.; Huang, Y.; Zhao, X.; Hu, Y.; Cai, Q.-N. Constitutive and Induced Activities of Defense-Related Enzymes in Aphid-Resistant and Aphid-Susceptible Cultivars of Wheat. *J. Chem. Ecol.* **2009**, *35*, 176–182. [CrossRef]
54. Aghdam, M.S.; Asghari, M.; Farmani, B.; Mohayeji, M.; Moradbeygi, H. Impact of postharvest brassinosteroids treatment on PAL activity in tomato fruit in response to chilling stress. *Sci. Hortic.* **2012**, *144*, 116–120. [CrossRef]
55. Zhan, L.; Hu, J.; Zhu, Z. Shelf life extension of minimally processed water caltrop (*Trapa acornis* Nakano) fruits coated with chitosan. *Int. J. Food Sci. Technol.* **2011**, *46*, 2634–2640. [CrossRef]
56. Kou, X.-H.; Guo, W.-L.; Guo, R.-Z.; Li, X.-Y.; Xue, Z.-H. Effects of Chitosan, Calcium Chloride, and Pullulan Coating Treatments on Antioxidant Activity in Pear cv. "Huang guan" During Storage. *Food Bioprocess Technol.* **2014**, *7*, 671–681. [CrossRef]
57. Lu, L.; Ji, L.; Shi, R.; Li, S.; Zhang, X.; Guo, Q.; Wang, C.; Qiao, L. Dextran as an elicitor of phenylpropanoid and flavonoid biosynthesis in tomato fruit against gray mold infection. *Carbohydr. Polym.* **2019**, *225*, 115236. [CrossRef] [PubMed]
58. Prasannath, K. Plant defense-related enzymes against pathogens: A review. *J. Agric. Sci.* **2017**, *11*, 38. [CrossRef]
59. Minh, N.P.; Nhi, T.T.Y.; Vien, L.T.B.; Ha, T.T.T.; Yen, N.T.K. Effect of Arabic Gum Coating on Postharvest Quality of Litchi (*Litchi chinensis*) Fruits. *J. Pharma Sci. Res.* **2019**, *11*, 1464–1468.
60. Eissa, H.A. Effect of chitosan coating on shelf life and quality of fresh-cut mushroom. *J. Food Qual.* **2007**, *30*, 623–645. [CrossRef]
61. Badawy, M.E.; Rabea, E.I. Potential of the biopolymer chitosan with different molecular weights to control postharvest gray mold of tomato fruit. *Postharvest Biol. Technol.* **2009**, *51*, 110–117. [CrossRef]
62. Kuvalekar, A.; Redkar, A.; Gandhe, K.; Harsulkar, A. Peroxidase and polyphenol oxidase activities in compatible host–pathogen interaction in *Jasminum officinaleand Uromyces hobsoni*: Insights into susceptibility of host. *N. Z. J. Bot.* **2011**, *49*, 351–359. [CrossRef]
63. Els, A.I.; Moha, A.H.; Ode, D.C.; Go, A.M. Biochemical Effects of Chitosan Coating and Hot Water Dipping on Green Bean Decay During Cold Storage. *J. Appl. Sci.* **2019**, *19*, 101–108. [CrossRef]
64. Adiletta, G.; Zampella, L.; Coletta, C.; Petriccione, M. Chitosan Coating to Preserve the Qualitative Traits and Improve Antioxidant System in Fresh Figs (*Ficus carica* L.). *Agriculture* **2019**, *9*, 84. [CrossRef]
65. Pasquariello, M.S.; Di Patre, D.; Mastrobuoni, F.; Zampella, L.; Scortichini, M.; Petriccione, M. Influence of postharvest chitosan treatment on enzymatic browning and antioxidant enzyme activity in sweet cherry fruit. *Postharvest Biol. Technol.* **2015**, *109*, 45–56. [CrossRef]

coatings

MDPI

Article

Novel Approach for Improving Papaya Fruit Storage with Carnauba Wax Nanoemulsion in Combination with *Syzigium aromaticum* and *Mentha spicata* Essential Oils

Josemar Gonçalves de Oliveira Filho [1,†], Larissa G. R. Duarte [2,†], Yasmin B. B. Silva [3], Eduardo P. Milan [2], Higor V. Santos [3], Thaís C. Moura [3], Vitor P. Bandini [3], Luís Eduardo S. Vitolano [3], Jacqueline J. C. Nobre [4], Cristiane T. Moreira [3], Milene C. Mitsuyuki [2], Stanislau Bogusz Junior [3] and Marcos David Ferreira [2,*]

[1] School of Pharmaceutical Sciences, São Paulo State University (UNESP), Araraquara 14800-903, Brazil; josemar.gooliver@gmail.com

[2] Brazilian Agricultural Research Corporation, Embrapa Instrumentation, São Carlos 13561-206, Brazil; larissagraziele@gmail.com (L.G.R.D.); edpmilan@gmail.com (E.P.M.); milene.corso@embrapa.br (M.C.M.)

[3] Institute of Chemistry (IQSC), University of São Paulo (USP), São Carlos 13566-590, Brazil; yasmin.belleze@gmail.com (Y.B.B.S.); higorsantos8@usp.br (H.V.S.); thaiscarvalho625@gmail.com (T.C.M.); vitor.bandini@usp.br (V.P.B.); luis.vitolano@usp.br (L.E.S.V.); cristiane.temporine@usp.br (C.T.M.); stanislau@iqsc.usp.br (S.B.J.)

[4] Department of Biotechnology, Federal University of São Carlos, São Carlos 13565-905, Brazil; jacqueline.cardoso@estudante.ufscar.br

* Correspondence: marcos.david@embrapa.br

† These authors contributed equally to this work.

Abstract: Application of hydrophobic coatings, such as carnauba wax nanoemulsions, combined with natural antimicrobials, has been demonstrated to be an effective solution in extending the shelf life of fruits. The present study evaluated the effectiveness of carnauba wax nanoemulsion (CWN) coatings containing free or encapsulated with β-cyclodextrin (β-CD) essential oils of *Syzigium aromaticum* (CEO) and *Mentha spicata* (MEO) for the post-harvest conservation of papaya fruit. The chemical composition of the essential oils (EOs) was analyzed using GC-MS. Subsequently, coatings incorporating free and encapsulated EOs were prepared and applied to papaya fruit. Fruit was evaluated for post-harvest quality parameters during 15 days of storage. Clove essential oil presented as main compounds eugenol (89.73%), spearmint and carvone (68.88%), and limonene (20.34%). The observed reduction in weight loss in coated fruit can be attributed to the formation of a physical barrier provided by the coating. Compared to the control group, which experienced the highest weight loss of 24.85%, fruit coated with CWN and CWN-MEO:β-CD exhibited significantly lower weight loss percentages of only 5.78% and 7.5%, respectively. Compared to the control group, which exhibited a release of ethylene at a rate of 1.3 $\mu g\ kg^{-1}\ h^{-1}$, fruit coated with CWN, CWN-MEO:β-CD, and CWN-MEO coatings demonstrated a lower ethylene release rate at 0.7 $\mu g\ kg^{-1}\ h^{-1}$. Although the physical-chemical properties of papayas, including pH, Brix, titratable acidity, color, and texture, remained largely unchanged during storage with the coatings, analysis of incidence and severity of papaya post-harvest deterioration revealed that coatings containing essential oils effectively acted as antifungals in the fruit. Microscopy images showed that CWN and CWN-MEO:β-CD coatings are more uniform compared to the others. The edible coatings, especially CWN and CWN-MEO: β-CD, can act as antimicrobial coatings on papaya fruit, increasing their conservation during post-harvest storage.

Keywords: natural antifungal compounds; post-harvest; preservative; hydrophobic coatings

Citation: Oliveira Filho, J.G.d.; Duarte, L.G.R.; Silva, Y.B.B.; Milan, E.P.; Santos, H.V.; Moura, T.C.; Bandini, V.P.; Vitolano, L.E.S.; Nobre, J.J.C.; Moreira, C.T.; et al. Novel Approach for Improving Papaya Fruit Storage with Carnauba Wax Nanoemulsion in Combination with *Syzigium aromaticum* and *Mentha spicata* Essential Oils. *Coatings* **2023**, *13*, 847. https://doi.org/10.3390/coatings13050847

Academic Editors: María B. Pérez-Gago and Lluís Palou

Received: 25 March 2023
Revised: 24 April 2023
Accepted: 25 April 2023
Published: 28 April 2023

1. Introduction

Currently there is a higher demand for healthy foods without synthetic preservatives by consumers [1]. Furthermore, foods rich in vitamins such as fruit and vegetables are highly perishable and susceptible to microbial deterioration [2–4]. Natural antimicrobial coatings have proven to be an excellent alternative to increase food shelf life [5,6].

Edible coatings based on lipids such as waxes and oils prevent the diffusion of water vapor and decrease the respiration rate because of their hydrophobic character [7,8]. The high rates of water loss and respiration in fruit lead to significant decreases in firmness, crispness, and weight resulting from biochemical changes that accelerate the process of deterioration [9,10].

Carnauba wax is extracted from the Brazilian palm tree *Copernicia prunifera* and is recognized as generally recognized as safe (GRAS) by the FDA, and its use is authorized by Anvisa, FAO-Food and Agriculture Organization of the United Nations, and the European Union [7,11,12].

Carnauba wax hydrophobicity is due to the high content of fatty alcohols and long-chain alkanes present in its structure [13], preventing water loss and also being described with antifungal action [14]. Nanoemulsions, with particle diameters ranging from 10 to 100 nm, possess higher clarity and translucency compared to conventional emulsions as their average size is smaller than the visible light wavelength ($r << \lambda$) [15,16]. Moreover, decreasing the particle diameter offers a promising strategy for generating more thermodynamically stable emulsions [17]. Previous studies have demonstrated that carnauba wax nanoemulsion has great potential as a fruit coating material, capable of prolonging their shelf life and imparting shine [7,12,18].

In order to impart antimicrobial properties to coatings, natural antimicrobial agents such as essential oils have been incorporated [12]. Essential oils have also been considered GRAS substances by the FDA since 2008 [19]. In addition to being hydrophobic, essential oils have antimicrobial activity [20,21] which may vary depending on the composition of the oil and species of microorganisms [22].

Syzigium aromaticum EO's main compound is eugenol, showing antimicrobial, antiviral and antioxidant activity [23,24]. On the other hand, for *Mentha spicata*, carvone is the main one, which is also an antioxidant, and due to this bioactivity, it increased shelf life of fresh meats [25]. The antimicrobial activity of carvone was proven in the study by [26] with the inhibition of the growth of the fungus *Colletotrichum gloeosporioides* in papayas and with antifungal action against *Botrytis cinerea* in plums [27].

Natural antimicrobials such as essential oils can be added directly to foods, acting as biopreservatives [28–30]. However, the effectiveness of these bioactive compounds can be impaired by their high volatility and they are prone to degradation caused by exposure to light, high temperatures, and the presence of oxygen [22,31]. In addition, essential oils have low solubility in water and a very intense aroma that can interfere with the sensory attributes of the food to which they are applied [32,33]. Encapsulation of essential oils offers potential solutions to overcome several challenges, including enhanced stability and protection, better control over compound release, reduced intensity of flavors and odors, prolonged shelf life, and improved bioavailability and palatability of the encapsulated materials [34].

Several studies have investigated the antibacterial properties of green mint nanoemulsions [35], while others have explored the antifungal potential of essential oil nanoemulsions containing thymol and were incorporated into quinoa and chitosan films [36]. However, there is a scarcity of research focused on nanoemulsions containing encapsulated clove and mint oils despite their notable antifungal activity and potential to provide a safer and more natural alternative to conventional antifungal agents.

The primary objective of this study was to assess the effectiveness of edible coatings composed of nanoemulsions containing carnauba wax, *S. aromaticum*, and *M. spicata* essential oils in preserving papayas. A post-harvest quality evaluation was conducted to determine the impact of free and encapsulated essential oils in the coatings on the retardation of fruit ripening and deterioration.

2. Materials and Methods

2.1. Materials

A sample of Carnauba wax type I with 99% purity and CAS No.: 8015-86-9 was obtained from Pontes Indústria de Cera in Fortaleza, CE, Brazil. MEO and CEO were acquired from Laszlo Aromaterapia in Belo Horizonte, MG, Brazil. Papaya fruit of the cultivar THB from the solo group were transported from a commercial farm in Bahia State to the postharvest laboratory at Embrapa Instrumentação in São Carlos, SP, where they were sanitized using a specialized fruit detergent and chlorine dioxide. Only papayas that lacked standard defects, met size requirements, and were at stage 1 of maturation (with less than 15% of their skin surface covered in yellow) were selected for the study [37].

2.2. Essential Oil Composition

A qualitative analysis of essential oils (EOs) was conducted via gas chromatography using a Shimadzu (GC-2010 Plus, Kyoto, Japan) coupled with a quadrupole mass spectrometer (GC-MS). A non-polar DB-5MS capillary column (30 m × 0.25 mm, i.d. × 0.25 μm) was used for gas chromatography analyses with helium as the carrier gas at a flow rate of 1 mL/min. Essential oil samples were diluted in dichloromethane (10% *v*/*v*) and injected (1 μL) in a split mode (1:50). The chromatographic conditions were as follows: injector temperature: 220 °C, oven temperature: 60 to 240 °C at 3 °C/min; interface: 240 °C; ion source: +70 eV, *m*/*z*: 35–350. The linear temperature programmed retention index (RI) was calculated using an alkane solution (C_7–C_{30}). Identification of analytes was conducted by comparing the RI and mass spectra obtained from the sample with mass spectra and RI of the literature, with at least 85% similarity for the mass spectra and maximum variation in RI of ± 10. The identification of analytes was confirmed by co-injection of authentic standards whenever available. Semi-quantitative analysis of essential oils (% relative area) was performed using the flame ionization detector (GC-FID) in the same gas chromatography system. All qualitative and semi-quantitative analyses were performed in triplicate.

2.3. Encapsulation of Essential Oils with β-Cyclodextrin

MEO:β-CD and CEO:β-CD microcapsules were prepared by the co-precipitation method as reported [38]. The MEO:β-CD and CEO:β-CD ratios of 10:90 and 20:80 (% *w*/*w*), respectively, were selected as these ratios provides the maximum inclusion of MEO or CEO in β-CD according to previous tests. Obtained MEO:β-CD and CEO:β-CD microcapsules were stored in a desiccator at 25 °C until use.

2.4. Edible Coating Preparation

A carnauba wax nanoemulsion (CWN) was prepared using an oil phase and water phase via a high-pressure process with ammonia in a morpholine-free method adapted for this study [39] in a high-pressure process. The diameter size of the CWN obtained was 44.1 ± 7.6 nm with a narrow polydispersion index of 0.28 and a zeta potential of −43.8 mV as measured by the Zetasizer Nano ZS (Malvern Instruments Inc., Westborough, MA, USA) [40]. The incorporation of MEO and CEO free and microencapsulated, as antimicrobial agents, was done by mixing the 1.0% concentration with CWN in a high-speed mixer (UltraTurrax T25, IKA Werke GmbH & Co, Staufen, Alemanha) for 5 min at 5.000 rpm.

The coatings were applied to the fruit, which were randomly divided into 6 treatment groups as follows: CWN (9% solid phase in suspension), CWN (9%) with MEO (1%), CWN (9%) with CEO (1%), CWN (9%) with MEO:β-CD (1%), CWN (9%) with CEO:β-CD (1%), and non-treated fruit as a control. The coatings were applied manually by pouring 1 mL of the coating solution onto latex-gloved hands and then manually spreading it on sanitized papayas. For non-destructive analyses, five papayas were used per treatment, and for destructive analyses, ten papayas were used. The fruit was stored for 15 days at 16 °C and a relative humidity of 70%. The quality attributes of the papayas were evaluated at the beginning of the experiment (0 days) and after 5, 10, and 15 days of storage.

2.5. Physicochemical Parameters of Papayas

The fruit weight loss was determined using the [41] standard method by measuring the fruit weight on day 0 (start of the experiment) and on days 5, 10, and 15 of storage. The percentage difference between the initial and final weight on each day was used to calculate the weight loss.

The soluble solids (SS) content was measured with an Atago RX-5000cx digital refractometer (Tokyo, Japan) and expressed as Brix following the [41] standard method. The pH of the samples was assessed using a PHS-3B digital pH meter following the same standard method. The titratable acidity was determined using 0.1 N NaOH and phenolphthalein as an indicator, and the results were expressed as g of citric acid per 100 g of fruit.

The color measurements were performed on the external surface of the fruit (on the peel) with a Konica Minolta CR-400 colorimeter (Konica Minolta, Osaka, Japan) equipped with a C illuminant using the CIELAB scale. Hue angle ($h°$), chroma (C^*), and total color difference (ΔE) were calculated with Equations (1)–(3), respectively.

$$h^{\circ} = tan^{-1}\left(\frac{b^*}{a^*}\right) \tag{1}$$

$$C^* = \left((a^*)^2 + (b^*)^2\right)^{1/2} \tag{2}$$

$$\Delta E^* = \sqrt{(L_t{}^* - L_{t0}{}^*)^2 + (a_t{}^* - a_{t0}{}^*)^2 + (b_t{}^* - b_{t0}{}^*)^2} \tag{3}$$

where subscripts t and 0 correspond to parameters evaluated at time t and at the beginning of the study, respectively.

The firmness of the fruit was assessed using a digital TA.XTplus Texture Analyzer (Stable Micro Systems Ltd., Godalming, UK) equipped with a 6 mm diameter probe, 15 mm/s velocity, 5 mm penetration distance, and 12 mm^2 contact area with the peel removed. The results were reported in Newtons (N) and the mean value was calculated based on three penetrations in the distal region of each fruit. All analyses were performed in triplicate and the data were presented as mean ± standard deviation.

2.6. Respiration Rate and Ethylene of Papayas

The respiration rate was determined following the method described by [42], using a respirometer (model 6600, Illinois Instrument, Inc., Johnsburg, IL, USA). Two papayas were placed in 2000 mL glass containers with a silicone septum in the lids, which were hermetically sealed. The concentrations of O_2 and CO_2 were measured at each time point by suctioning air samples from the containers using a paramagnetic sensor and an infrared sensor, respectively. Ethylene production was determined according to the method described by [18]. Two papayas of the same treatment were packed in pairs in hermetic glass jars with screw caps and held for 2 h. At the end of this period, 1 mL of the headspace was collected through a rubber septum located on the cap. This volume was injected with Varian Gas Chromatograph model CP 3800, with TCD/FID detectors, in order to detect the peaks corresponding to ethylene. Results were expressed in $\mu g \cdot kg^{-1} \cdot h^{-1}$. All analyses were carried out in triplicate, and the data were calculated as means ± standard deviations.

2.7. Scanning Electron Microscopy

Images of papaya peels with or without coating were determined according to [43] by emission gun scanning electron microscopy (SEM-SEM JEOL JSM-6701F, Tokyo, Japan). Surface and fracture micrographs of the fruit peel were obtained. Both were first dried and then coated with gold. The accelerating voltage used for microscopy was 10 kV.

2.8. Decay Percentage and Severity on Papayas

The presence or absence of mold growth in papayas during storage was evaluated visually, and any visible spoilage was considered as decay. The percentage of decay was determined based on the number of decayed papayas per treatment, with each treatment having ten papayas. The severity of the disease in the fruit was assessed using a six-point scale (0 = no symptoms; 1 = 1%–20% affected area; 2 = 21%–40%; 3 = 41%–60%; 4 = 61%–80%; and 5 = 81%–100%) and was used to evaluate the antifungal activity of the treatments [12].

2.9. Statistical Analysis

Data were described using means and standard deviations and comparison of means was performed by parametric analysis of variance and Duncan's multiple comparisons test or non-parametric ANOVA and Kruskal–Wallis multiple comparisons test, depending on the homogeneity condition of variance, verified by the Bartlett test, or the level of measurement of the response variable. The significance level was set at 5% and the software used for the analyses was R version 4.2.2.

3. Results and Discussion

3.1. Essential Oil Composition

The major compounds of clove (*Syzigium aromaticum*) and spearmint (*Mentha spicata*) essential oils obtained from chromatograms are shown in Table 1. Clove essential oil presented as main compounds eugenol (89.73%), spearmint and carvone (68.88%), and limonene (20.34%). These results are close to those found by [44], who obtained eugenol values (70.58%), and [45], who found 96.33% eugenol for clove oil. Reference [46] obtained 62.9% carvone and 8.5% limonene for spearmint oil. The variation in chemical composition in the comparison of the mentioned works may be due to factors such as geographic origin, environmental conditions, age and part of the plant, seasonal and climatic conditions, genetic factors, and even plant nutrition [47,48].

Table 1. Composition of essential oils.

Compound	*Syzigium aromaticum* (% Area)	*Mentha spicata* (% Area)
α-Pinene	-	0.69
Sabinene	-	0.32
β-Pinene	-	0.76
Myrcene	-	0.95
3-Octanol	-	0.25
p-Cymene	-	0.23
Limonene	-	20.34
1,8-Cineol	-	1.10
γ-Terpinene	-	0.13
Menthone	-	0.50
cis-Sabinene hydrate	-	0.17
Menthol	-	0.15
Isomenthol	-	1.06
(E)-dihydrocarvone	-	1.40
cis-Dihydrocarvone	-	0.15
trans-Carveol	-	0.28
Carvone	-	68.88
Piperitone	-	0.18
Menthyl acetate	-	0.39
Dihydrocarvyl acetate	-	0.13
Eugenol	89.73	-
cis-Carvyl acetate	-	0.11
β-bourbenene	-	0.77
β-Gurjenene	7.59	-
Caryophyllene	-	1.03
α-Humulene	2.10	-
γ-Selinene	0.20	-
δ-Cadinene	0.25	-
Caryophyllene oxide	0.13	-
Total	100	99.97

Eugenol, present in clove essential oil, is a phytochemical that confers antimicrobial and antioxidant properties in addition to the characteristic flavor and odor of this oil [49–51].

The antifungal efficacy of carvone present in spearmint essential oil has been proven in other studies [52,53].

3.2. *Physicochemical Parameters of Papayas*

Based on the results presented in Figure 1, the control group exhibited the greatest weight loss (24.85%) after 15 days of storage, which was significantly different from all other treatments. In contrast, papayas treated with only carnauba wax nanoemulsion demonstrated the lowest weight loss (5.78%), followed by CWN-MEO:β-CD, which exhibited 7.5% weight loss. Comparison of the treatments containing essential oils revealed that papayas treated with carnauba nanoemulsion containing MEO, either free or encapsulated, exhibited lower weight loss, particularly during the first 10 days of storage, compared to those treated with CEO.

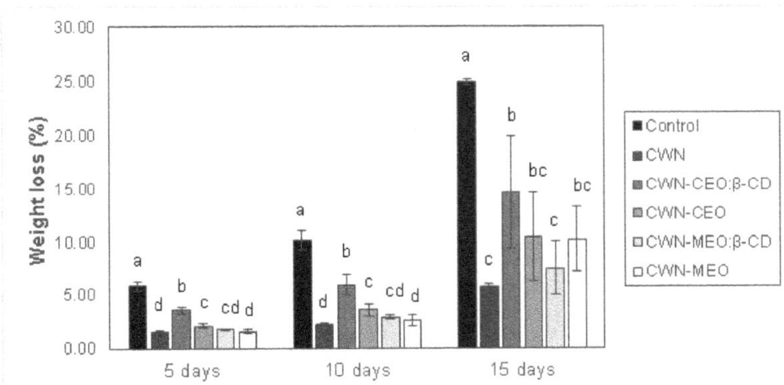

Figure 1. Papaya weight loss during storage time at 16 °C and 70% RH. For each storage period, different letters indicate significant differences between treatments ($p < 0.05$).

Fruit weight loss occurs through transpiration due to the respiration that takes place in the stomata of the epidermis [54,55]. The main component of the coating, carnauba wax, is highly hydrophobic; therefore, the coating acted as a barrier to gas exchange and thus reduced the transpiration rate of the fruit [7,56]. Similar results were obtained in cucumbers [57], apples [58], and papayas [12].

Ideal coatings should allow controlled gas exchange, avoiding the formation of anaerobic conditions and the accumulation of undesirable compounds, such as acetaldehydes and other off-flavors [59]. Since the weight loss of all treatments involving carnauba wax nanoemulsion with or without essential oils differed significantly from that of the control, the findings suggest that the coatings created a physical barrier that reduced the extent of fruit weight loss.

The results of pH, titratable acidity, and soluble solids analysis of papayas are presented in Table 2. The pH values of all treatments increased over time and were not significantly different among them (Table 2). A delay in the fruit ripening process occurs when there is a decrease in the use of some organic acids that are converted into sugars [12,60].

The increase in TSS values is directly related to the ripening of the fruit; as time passes, starch hydrolysis occurs and consequently the synthesis of sucrose and hexose in plant tissues [28,61]. The increase in soluble solids content is also attributed to a reduction in the water content of the fruit, resulting in a higher concentration of soluble solids [62,63].

Over time, the TA values of fruit tend to decrease as organic acids such as citric acid are used up during respiration [64,65]. Even though there were no significant differences in the pH, TA, and TSS values, the weight loss results indicate that the coating process inhibited the fruit's respiratory system.

Table 3 shows the values of the color parameters L^*, C^*, h°, and ΔE^*. The luminosity values (L^*) decreased over time for all treatments, indicating fruit ripening [66], with the exception of the group with encapsulated spearmint essential oils (CWN-MEO:β-CD), which showed higher values compared to the first day. The chroma (C^*) values of all groups also decreased, especially after the 5th day of storage, due to oxidative phenomena and the synthesis of papaya pigments such as carotene, lycopene, and anthocyanins during storage [67]. At the end of 15 days, the h° values of all treatments also decreased; as papaya matures, its color changes from greenish to yellowish due to chlorophyll degradation and carotenoid biosynthesis [68]. However, the CWN-MEO:β-CD treatment showed a statistical difference compared to the other treatments in relation to h° values. These changes in color parameters led to an increase in the total color difference (ΔE) of the fruits during storage, highlighting differences in h° values compared to day 0 (Table 3). By the end of 15 days, the fruits showed significantly equal ΔE values, indicating visible similarity among treatments.

During storage, fruit firmness decreased over time for all groups (as shown in Table 4). On the last day of storage (15th day), CWN-CEO showed the lowest reduction in firmness, while CWN-MEO showed the highest reduction (Table 4). The lowest firmness values observed for CWN-MEO may be the result of the interaction of the components of this oil with the cellular tissue of the fruit, causing structural changes that lead to softening and an increased release of enzymes or substrates that favor this process [69,70]. Some essential oils, depending on the concentration, can penetrate the cell tissue of the fruit and cause structural changes, decreasing firmness [6]. A similar behavior was reported by [71] for fresh-cut melons with alginate-based coatings that contained geraniol.

Table 2. Titratable acidity (TA), soluble solid content (TSS), and pH of fruit over time.

Treatments	pH			TSS (%)			TA (%)		
Storage Time (Days)	5	10	15	5	10	15	5	10	15
Control	5.35 ± 0.14 ^{b}	5.68 ± 0.10 ^{b}	6.05 ± 0.19 ^{ab}	10.02 ± 1.44 ^{a}	8.06 ± 1.45 ^{b}	8.97 ± 2.31 ^{a}	0.066 ± 0.013 ^{bc}	0.097 ± 0.090 ^{a}	0.055 ± 0.014 ^{a}
CWN	5.64 ± 0.05 ^{a}	6.00 ± 0.13 ^{a}	6.28 ± 0.23 ^{a}	8.59 ± 1.51 ^{a}	8.87 ± 1.19 ^{ab}	7.75 ± 1.32 ^{a}	0.065 ± 0.009 ^{c}	0.052 ± 0.007 ^{a}	0.059 ± 0.010 ^{a}
CWN-CEO:β-CD	5.48 ± 0.11 ^{ab}	5.69 ± 0.14 ^{b}	6.21 ± 0.14 ^{ab}	8.76 ± 0.76 ^{a}	9.43 ± 1.40 ^{ab}	9.27 ± 1.99 ^{a}	0.070 ± 0.010 ^{bc}	0.057 ± 0.005 ^{a}	0.051 ± 0.009 ^{a}
CWN-CEO	5.49 ± 0.08 ^{ab}	5.61 ± 0.21 ^{b}	6.09 ± 0.12 ^{ab}	8.80 ± 0.87 ^{a}	9.02 ± 0.67 ^{ab}	7.43 ± 1.20 ^{a}	0.085 ± 0.012 ^{a}	0.056 ± 0.005 ^{a}	0.050 ± 0.008 ^{a}
CWN-MEO:β-CD	5.45 ± 0.10 ^{ab}	5.54 ± 0.14 ^{b}	6.11 ± 0.10 ^{ab}	9.02 ± 0.70 ^{a}	8.67 ± 0.81 ^{b}	7.90 ± 1.11 ^{a}	0.081 ± 0.009 ^{ab}	0.051 ± 0.006 ^{a}	0.059 ± 0.006 ^{a}
CWN-MEO	5.46 ± 0.27 ^{ab}	5.72 ± 0.21 ^{b}	6.00 ± 0.24 ^{b}	9.84 ± 1.24 ^{a}	10.37 ± 1.17 ^{a}	8.25 ± 1.90 ^{a}	0.068 ± 0.004 ^{bc}	0.048 ± 0.011 ^{a}	0.056 ± 0.020 ^{a}

Means followed by different letters on the same column indicate significant differences between treatments ($p < 0.05$).

Table 3. Color parameters L*, C*, h°, and ΔE of papayas stored for 15 days at 16 °C and 70% RH.

Treatments	Time (Days)														
	0			5				10				15			
	*L**	*C**	(*h*°)	*L**	*C**	(*h*°)	Δ*E**	*L**	*C**	(*h*°)	Δ*E**	*L**	*C**	(*h*°)	Δ*E**
Control	54.85 ± 3.91 ^{a}	39.00 ± 2.13 ^{a}	102.92 ± 2.55 ^{a}	61.48 ± 5.77 ^{a}	49.53 ± 6.98 ^{a}	101.74 ± 9.47 ^{a}	13.73 ± 5.74 ^{a}	53.76 ± 6.93 ^{a}	27.00 ± 4.31 ^{a}	88.83 ± 11.20 ^{a}	16.60 ± 1.54 ^{a}	47.52 ± 10.88 ^{b}	22.55 ± 7.14 ^{b}	74.59 ± 12.82 ^{a}	24.31 ± 6.82 ^{a}
CWN	52.00 ± 3.39 ^{b}	37.43 ± 1.89 ^{a}	103.70 ± 2.34 ^{a}	49.37 ± 4.75 ^{c}	23.73 ± 3.05 ^{b}	102.44 ± 7.42 ^{a}	14.55 ± 3.05 ^{a}	52.53 ± 6.55 ^{a}	25.92 ± 3.06 ^{a}	96.70 ± 10.69 ^{a}	14.32 ± 2.41 ^{a}	50.84 ± 8.58 ^{b}	24.85 ± 4.75 ^{b}	85.66 ± 15.19 ^{a}	18.45 ± 3.33 ^{a}
CWN-CEO:β-CD	53.10 ± 3.40 ^{ab}	37.63 ± 2.04 ^{a}	104.08 ± 2.33 ^{a}	53.28 ± 7.65 ^{bc}	31.79 ± 11.88 ^{b}	103.79 ± 9.23 ^{a}	13.14 ± 6.30 ^{a}	54.43 ± 6.51 ^{a}	27.80 ± 3.47 ^{a}	92.87 ± 10.63 ^{a}	14.15 ± 2.78 ^{a}	49.83 ± 10.22 ^{b}	24.58 ± 4.24 ^{b}	80.35 ± 14.02 ^{a}	20.19 ± 3.78 ^{a}
CWN-CEO	54.86 ± 4.94 ^{a}	38.45 ± 3.75 ^{a}	102.89 ± 2.97 ^{a}	61.49 ± 4.76 ^{a}	48.17 ± 5.11 ^{a}	103.86 ± 5.61 ^{a}	12.32 ± 5.16 ^{a}	56.55 ± 6.08 ^{a}	28.43 ± 3.84 ^{a}	87.93 ± 8.81 ^{a}	15.15 ± 2.54 ^{a}	52.44 ± 10.35 ^{ab}	26.21 ± 6.18 ^{b}	74.64 ± 10.14 ^{a}	21.59 ± 3.42 ^{a}
CWN-MEO:β-CD	53.63 ± 4.10 ^{ab}	37.11 ± 2.60 ^{a}	103.82 ± 2.66 ^{a}	53.96 ± 5.89 ^{bc}	26.25 ± 3.31 ^{b}	97.05 ± 30.40 ^{ab}	12.17 ± 1.27 ^{a}	58.54 ± 5.65 ^{a}	30.05 ± 3.48 ^{a}	86.57 ± 27.61 ^{a}	14.17 ± 3.51 ^{a}	58.01 ± 7.39 ^{a}	31.75 ± 4.56 ^{a}	71.07 ± 29.77 ^{a}	21.28 ± 3.73 ^{a}
CWN-MEO	55.16 ± 4.21 ^{a}	39.71 ± 4.25 ^{a}	101.21 ± 4.05 ^{b}	54.38 ± 5.80 ^{b}	26.69 ± 3.84 ^{b}	94.43 ± 9.61 ^{b}	14.48 ± 2.01 ^{a}	57.84 ± 5.97 ^{a}	29.38 ± 4.21 ^{a}	87.37 ± 10.81 ^{a}	15.03 ± 2.49 ^{a}	50.84 ± 8.58 ^{b}	24.85 ± 4.75 ^{b}	85.66 ± 15.19 ^{a}	20.63 ± 5.07 ^{a}

Means followed by different letters on the same column indicate significant differences between treatments ($p < 0.05$).

Table 4. Firmness (N) of papaya during storage for 15 days at 16 °C and 70% RH.

Treatments	Storage Time (Days)		
	5	10	15
Control	5.46 ± 2.22 ^{a}	5.52 ± 3.07 ^{a}	4.02 ± 3.07 ^{ab}
CWN	4.32 ± 3.02 ^{a}	3.68 ± 1.33 ^{a}	4.68 ± 2.20 ^{ab}
CWN-CEO:β-CD	5.53 ± 3.10 ^{a}	4.34 ± 0.92 ^{a}	3.87 ± 2.30 ^{ab}
CWN-CEO	6.59 ± 3.79 ^{a}	3.83 ± 1.04 ^{a}	6.60 ± 2.41 ^{a}
CWN-MEO:β-CD	5.65 ± 3.35 ^{a}	7.12 ± 2.10 ^{a}	4.71 ± 1.70 ^{ab}
CWN-MEO	5.15 ± 2.98 ^{a}	3.88 ± 0.32 ^{a}	2.82 ± 0.58 ^{b}

Means followed by different letters on the same column indicate significant differences between treatments ($p < 0.05$).

The effectiveness of the carnauba wax nanoemulsion coating without and with essential oils in ethylene release, CO_2 production rate, and O_2 consumption rate of papayas can be seen in Figure 2. Applying coatings to papayas during storage resulted in a reduction in the ethylene production of the fruit, ultimately leading to delayed maturation [72]. No significant difference in ethylene levels during storage was observed among the different coatings. Ethylene is a hormone related to fruit ripening, and high levels indicate fast ripening [73]. Similar behavior was observed by [74] for plums coated with hydroxypropylmethylcellulose and two different essential oils (oregano essential oil (OEO) and bergamot essential oil (BEO)), fruit coated with H-OEO showed no significant difference in the production of ethylene compared to fruit coated with H-BEO, and both oils were effective in reducing and delaying ethylene production.

The levels of CO_2 and O_2 were also measured during storage and are shown in Figure 2. The control group showed a significant increase in CO_2 production (Figure 2B) and O_2 consumption (Figure 2C), indicating high metabolism and accelerated ripening, which ultimately led to a shorter shelf life. Regarding the coating treatments, it is worth noting that fruit coated with CWN exhibited lower CO_2 production after 5, 10, and 15 days of storage, as well as lower O_2 consumption.

The balance between those two gases enhances post-harvest life. High levels of CO_2 in the fruit restrict the Krebs cycle and low levels of O_2 inhibit the activities of respiratory enzymes [75]. Association with OEs did not show a significant decrease in CO_2 or in O_2 at 5 and 10 days of storage. It can be seen that the CWN-CEO coating presented O_2 concentrations close to that of the control (Figure 2C) as coatings can present different degrees of permeability due to the formation of irregular structures and thicknesses during film consolidation [41]. However, at 15 days of storage, the treatments wit OE encapsulated demonstrated a reduction on O_2 consumption when compared to control, a possible indication of reduction in metabolism.

3.3. Scanning Electron Microscopy

For a better understanding of the deposition of the coatings on the fruit, microscopic analyses were carried out as shown in Figure 3 (micrographs of the surface of the peels and micrographs of the fractures of the peels). Microscopic analysis showed that the CWN and CWN-MEO:β-CD coatings were more uniformly applied over the fruit surface compared to the other treatments, which exhibited more cleavage or cracking. This result is consistent with the findings for the CWN coating, which showed the lowest weight loss (Figure 1) and the highest inhibition of ethylene biosynthesis and gas exchange (CO_2 and O_2) (Figure 2) in papayas, followed by the CWN-MEO:β-CD coating.

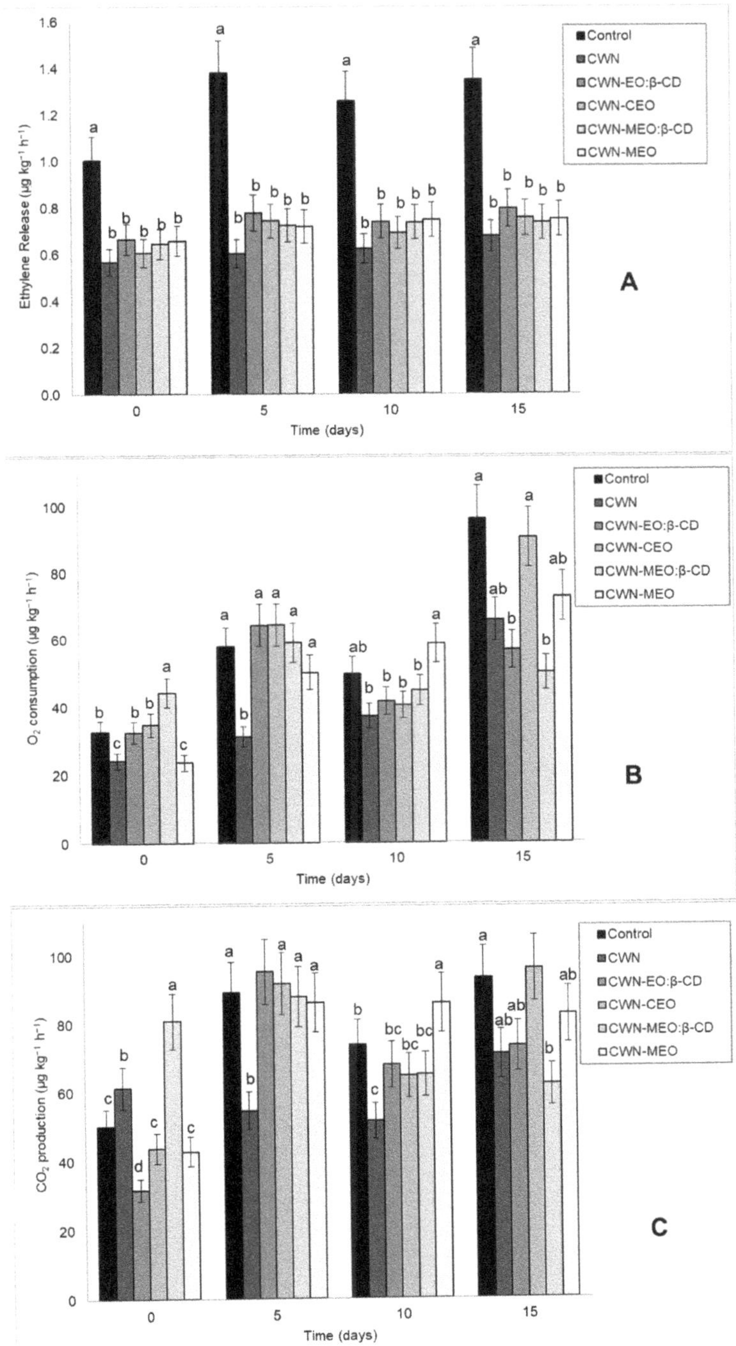

Figure 2. (**A**) Ethylene release, (**B**) CO_2 production rate, and (**C**) O_2 consumption rate of papayas during storage for 15 days at 16 °C and 70% RH. For each storage period, different letters indicate significant differences between treatments ($p < 0.05$).

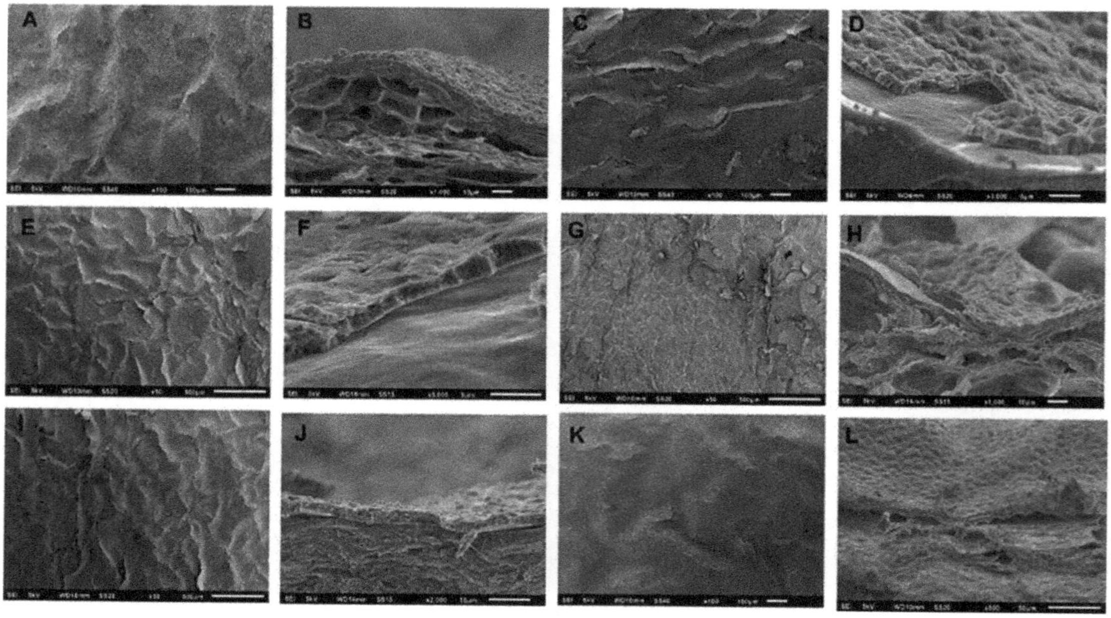

Figure 3. Micrographs of the surface (magnification 50× and scale bar 500 μm) and fractures (magnification 2000× and scale bar 10 μm) of the peels of papaya Control (**A**,**B**), CWN (**C**,**D**) CWN-CEO:β-CD (**E**,**F**) CWN-CEO (**G**,**H**) CWN-MEO:β-CD (**I**,**J**) and CWN-MEO (**K**,**L**).

The greater the chemical homogeneity of the nanometric coatings, the greater the uniformity and adhesion to the fruit [76]. CWN presents a regular structure due to the stability of the nanoemulsion (with the diameter size parameters of 44, 1 nm, PDI 0.28, and zeta potential of −43.8 mV) [38]. The uniformity of CWN-MEO:β-CD is due to encapsulation avoiding aggregation and flocculation of EO droplets ensuring a better distribution of EO in coatings [77]. However, the chemical composition of the EO determines its polarity and viscosity; thus, the type of EO can affect the average droplet size of the nanoemulsion [78], so CWN-CEO:β-CD may not have shown as much uniformity compared to CWN-MEO:β-CD (Figure 3).

3.4. Decay Percentage and Severity on Papayas

Coatings reduced postharvest disease incidence (Figure 4A) and severity (Figure 4B–D) in papayas when compared to control fruit. The CWN-MEO:β-CD coating showed the lowest incidence of disease compared to the other treatments at the end of 15 days. Encapsulated essential oils show greater stability in vivo tests due to the slow release of active compounds from the EO, reducing fruit rot in the long term. This behavior was also described by [79], who developed polylactic acid (PLA) nanocapsules with lemongrass EO and evaluated in vivo against the postharvest activity of *C. gloeosporioides* in apples.

Coatings with CWN-CEO and CWN-MEO essential oils had the lowest rot severities with 100% and 90% scores of 1%–20% affected area, respectively, at the end of the storage period. *S. aromaticum* and *M. spicata* essential oils added to CWN acted as antifungals. The antifungal mechanism of essential oils is through depolarization of the mitochondrial membrane and consequently greater cell permeability and imbalance in ion transport and thus cell death by apoptosis [80]. The antifungal action of the oils delayed fruit rot.

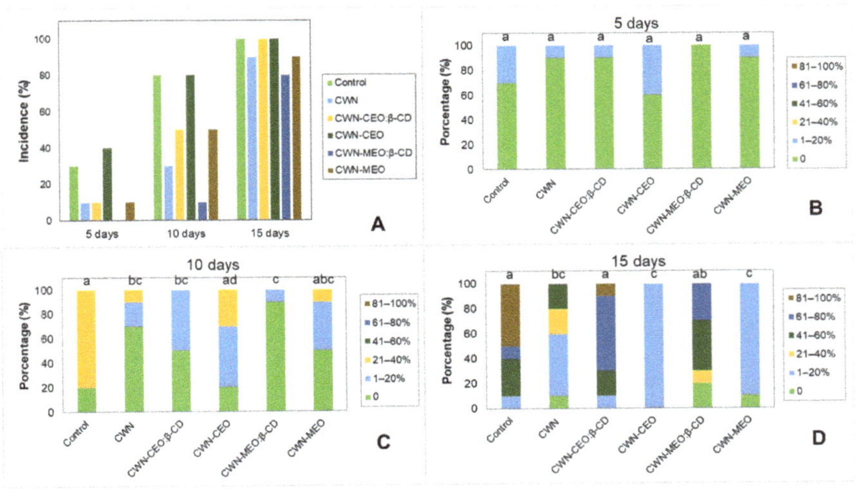

Figure 4. Fruit incidence (**A**) and severity of papaya post-harvest deterioration after 5 days (**B**), 10 days (**C**), and 15 days (**D**) of storage at 16 °C. For each storage period, different letters indicate significant differences between treatments ($p < 0.05$).

4. Conclusions

CWN coatings with or without essential oils reduced weight loss and delayed fruit rot due to physical barrier on gas exchange and presence of antifungal compounds. The microscopy images indicated that the CWN and CWN-MEO:β-CD coatings exhibited more uniformity and improved stability resulting from encapsulation in spearmint oil. The coating based on CWN-MEO was less effective in reducing fruit firmness loss due to negative interactions between MEO components and fruit tissue. The CWN-CEO-based coating was also inefficient in reducing the respiration rate of the fruit. Additionally, this coating did not show good uniformity when applied to papaya fruit, as observed in SEM images. This lack of uniformity negatively impacted gas exchange reduction, resulting in low coating efficiency. Coatings with carnauba nanoemulsion and essential oils inhibited the growth of fungi evaluated by the incidence and severity in the fruit. Therefore, CWN coatings with essential oils delayed fruit rot and thus can be a good alternative for natural antifungals and fruit preservation.

Author Contributions: Conceptualization, J.G.d.O.F. and M.D.F.; methodology, J.G.d.O.F., S.B.J. and M.D.F.; software, M.C.M.; formal analysis, L.G.R.D., Y.B.B.S. and M.C.M.; investigation, Y.B.B.S., E.P.M., H.V.S., T.C.M., V.P.B., L.E.S.V., J.J.C.N. and C.T.M.; resources, S.B.J. and M.D.F.; data curation, J.G.d.O.F.; writing—original draft preparation, J.G.d.O.F. and L.G.R.D.; writing—review and editing, M.D.F.; visualization, J.G.d.O.F. and L.G.R.D.; supervision, S.B.J. and M.D.F.; project administration, M.D.F. and J.G.d.O.F.; funding acquisition, M.D.F. All authors have read and agreed to the published version of the manuscript.

Funding: This research was funded by Embrapa (process 20.19.03.024.00.00), FAPESP (process 2022/10686-6), CAPES (Grant No. 001), and the Brazilian National Council for Scientific and Technological Development-CNPq (process 381280/2023-3).

Institutional Review Board Statement: Not applicable.

Informed Consent Statement: Not applicable.

Data Availability Statement: The authors confirm that the data supporting the findings of this study are available within the article.

Acknowledgments: The authors thank Empresa Brasileira de Pesquisa Agropecuária–Embrapa, Rede Agronano, and CNPq/MCTI Sisnano (process 442575/2019-0) from Brazil. Ferreira thanks CNPq for his Research Productivity fellowship (process 307141/2022-5).

Conflicts of Interest: The authors declare no conflict of interest.

References

1. Mesías, F.J.; Martín, A.; Hernández, A. Consumers' growing appetite for natural foods: Perceptions towards the use of natural preservatives in fresh fruit. *Food Res. Int.* **2021**, *150*, 110749. [CrossRef] [PubMed]
2. Alegbeleye, O.; Odeyemi, O.A.; Strateva, M.; Stratev, D. Microbial spoilage of vegetables, fruits and cereals. *Appl. Food Res.* **2022**, *2*, 100122. [CrossRef]
3. Duan, C.; Meng, X.; Meng, J.; Khan, M.I.H.; Dai, L.; Khan, A.; An, X.; Zhang, J.; Huq, T.; Ni, Y. Chitosan as A Preservative for Fruits and Vegetables: A Review on Chemistry and Antimicrobial Properties. *J. Bioresour. Bioprod.* **2019**, *4*, 11–21. [CrossRef]
4. Perumal, A.B.; Huang, L.; Nambiar, R.B.; He, Y.; Li, X.; Sellamuthu, P.S. Application of essential oils in packaging films for the preservation of fruits and vegetables: A review. *Food Chem.* **2022**, *375*, 131810. [CrossRef] [PubMed]
5. Basumatary, I.B.; Mukherjee, A.; Katiyar, V.; Kumar, S. Biopolymer-based nanocomposite films and coatings: Recent advances in shelf-life improvement of fruits and vegetables. *Crit. Rev. Food Sci. Nutr.* **2020**, *62*, 1912–1935. [CrossRef] [PubMed]
6. Yousuf, B.; Wu, S.; Siddiqui, M.W. Incorporating essential oils or compounds derived thereof into edible coatings: Effect on quality and shelf life of fresh/fresh-cut produce. *Trends Food Sci. Technol.* **2021**, *108*, 245–257. [CrossRef]
7. Miranda, M.; Ribeiro, M.D.M.M.; Spricigo, P.C.; Pilon, L.; Mitsuyuki, M.C.; Correa, D.S.; Ferreira, M.D. Carnauba wax nanoemulsion applied as an edible coating on fresh tomato for postharvest quality evaluation. *Heliyon* **2022**, *8*, e09803. [CrossRef]
8. Yousuf, B.; Sun, Y.; Wu, S. Lipid and Lipid-containing Composite Edible Coatings and Films. *Food Rev. Int.* **2021**, *38*, 574–597. [CrossRef]
9. Nawab, A.; Alam, F.; Hasnain, A. Mango kernel starch as a novel edible coating for enhancing shelf- life of tomato (*Solanum lycopersicum*) fruit. *Int. J. Biol. Macromol.* **2017**, *103*, 581–586. [CrossRef]
10. Susmita Devi, L.; Kalita, S.; Mukherjee, A.; Kumar, S. Carnauba wax-based composite films and coatings: Recent advancement in prolonging postharvest shelf-life of fruits and vegetables. *Trends Food Sci. Technol.* **2022**, *129*, 296–305. [CrossRef]
11. de Freitas, C.A.S.; de Sousa, P.H.M.; Soares, D.J.; da Silva, J.Y.G.; Benjamin, S.R.; Guedes, M.I.F. Carnauba wax uses in food—A review. *Food Chem.* **2019**, *291*, 38–48. [CrossRef]
12. Filho, J.G.D.O.; Silva, G.D.C.; Oldoni, F.C.A.; Miranda, M.; Florencio, C.; de Oliveira, R.M.D.; Gomes, M.D.P.; Ferreira, M.D. Edible Coating Based on Carnauba Wax Nanoemulsion and Cymbopogon martinii Essential Oil on Papaya Postharvest Preservation. *Coatings* **2022**, *12*, 1700. [CrossRef]
13. de Oliveira Filho, J.G.; Bezerra, C.C.D.O.N.; Albiero, B.R.; Oldoni, F.C.A.; Miranda, M.; Egea, M.B.; de Azeredo, H.M.C.; Ferreira, M.D. New approach in the development of edible films: The use of carnauba wax micro- or nanoemulsions in arrowroot starch-based films. *Food Packag. Shelf Life* **2020**, *26*, 100589. [CrossRef]
14. de Castro e Silva, P.; de Oliveira, A.C.S.; Pereira, L.A.S.; Valquíria, M.; Carvalho, G.R.; Miranda, K.W.E.; Marconcini, J.M.; Oliveira, J.E. Development of bionanocomposites of pectin and nanoemulsions of carnauba wax and neem oil pectin/carnauba wax/neem oil composites. *Polym. Compos.* **2020**, *41*, 858–870. [CrossRef]
15. Mayer, S.; Weiss, J.; McClements, D.J. Vitamin E-enriched nanoemulsions formed by emulsion phase inversion: Factors influencing droplet size and stability. *J. Colloid Interface Sci.* **2013**, *402*, 122–130. [CrossRef] [PubMed]
16. McClements, D.J.; Rao, J. Food-Grade Nanoemulsions: Formulation, Fabrication, Properties, Performance, Biological Fate, and Potential Toxicity. *Crit. Rev. Food Sci. Nutr.* **2011**, *51*, 285–330. [CrossRef]
17. McClements, D.J. Edible nanoemulsions: Fabrication, properties, and functional performance. *Soft Matter* **2011**, *7*, 2297–2316. [CrossRef]
18. Zucchini, N.M.; Florencio, C.; Miranda, M.; Borba, K.R.; Oldoni, F.C.A.; Oliveira Filho, J.G.; Bonfim, N.S.; Rodrigues, K.A.; de Oliveira, R.M.D.; Mitsuyuki, M.C.; et al. Effect of carnauba wax nanoemulsion coating on postharvest papaya quality. *Acta Hortic.* **2021**, *1325*, 199–206. [CrossRef]
19. Yan, J.; Wu, H.; Shi, F.; Wang, H.; Chen, K.; Feng, J.; Jia, W. Antifungal activity screening for mint and thyme essential oils against *Rhizopus stolonifer* and their application in postharvest preservation of strawberry and peach fruits. *J. Appl. Microbiol.* **2021**, *130*, 1993–2007. [CrossRef]
20. Chen, W.; Ma, S.; Wang, Q.; McClements, D.J.; Liu, X.; Ngai, T.; Liu, F. Fortification of edible films with bioactive agents: A review of their formation, properties, and application in food preservation. *Crit. Rev. Food Sci. Nutr.* **2021**, *62*, 5029–5055. [CrossRef]
21. Pandey, V.K.; Islam, R.U.; Shams, R.; Dar, A.H. A comprehensive review on the application of essential oils as bioactive compounds in Nano-emulsion based edible coatings of fruits and vegetables. *Appl. Food Res.* **2022**, *2*, 100042. [CrossRef]
22. Fonseca, J.D.M.; Trevisol, T.C.; Valencia, G.A.; Monteiro, A.R. Recent trends in composite nanoemulsions for food packaging applications. In *Bio-Based Nanoemulsions Agri-Food Applications*; Elsevier: Amsterdam, The Netherlands, 2022; pp. 387–398. [CrossRef]

23. Haro-González, J.N.; Castillo-Herrera, G.A.; Martínez-Velázquez, M.; Espinosa-Andrews, H. Clove Essential Oil (*Syzygium aromaticum* L. Myrtaceae): Extraction, Chemical Composition, Food Applications, and Essential Bioactivity for Human Health. *Molecules* **2021**, *26*, 6387. [CrossRef]
24. Wang, H.; Ma, Y.; Liu, L.; Liu, Y.; Niu, X. Incorporation of clove essential oil nanoemulsion in chitosan coating to control Burkholderia gladioli and improve postharvest quality of fresh Tremella fuciformis. *LWT* **2022**, *170*, 114059. [CrossRef]
25. Kanatt, S.R.; Chander, R.; Sharma, A. Antioxidant potential of mint (*Mentha spicata* L.) in radiation-processed lamb meat. *Food Chem.* **2007**, *100*, 451–458. [CrossRef]
26. Sarkhosh, A.; Schaffer, B.; Vargas, A.I.; Palmateer, A.J.; Lopez, P.; Soleymani, A.; Farzaneh, M. Antifungal activity of five plant-extracted essential oils against anthracnose in papaya fruit. *Biol. Agric. Hortic.* **2017**, *34*, 18–26. [CrossRef]
27. Aminifard, M.H.; Mohammadi, S. Essential oils to control Botrytis cinerea in vitro and in vivo on plum fruits. *J. Sci. Food Agric.* **2013**, *93*, 348–353. [CrossRef] [PubMed]
28. Duarte, L.G.R.; Ferreira, N.C.A.; Fiocco, A.C.T.R.; Picone, C.S.F. Lactoferrin-Chitosan-TPP Nanoparticles: Antibacterial Action and Extension of Strawberry Shelf-Life. *Food Bioprocess Technol.* **2023**, *16*, 135–148. [CrossRef]
29. Duarte, L.G.R. Natural antimicrobials: Healthy alternative of application in foods as preservatives. *Int. J. Health Sci.* **2022**, *2*. [CrossRef]
30. Bouarab Chibane, L.; Degraeve, P.; Ferhout, H.; Bouajila, J.; Oulahal, N. Plant antimicrobial polyphenols as potential natural food preservatives. *J. Sci. Food Agric.* **2019**, *99*, 1457–1474. [CrossRef]
31. Sumalan, R.M.; Kuganov, R.; Obistioiu, D.; Popescu, I.; Radulov, I.; Alexa, E.; Negrea, M.; Salimzoda, A.F.; Sumalan, R.L.; Cocan, I. Assessment of Mint, Basil, and Lavender Essential Oil Vapor-Phase in Antifungal Protection and Lemon Fruit Quality. *Molecules* **2020**, *25*, 1831. [CrossRef] [PubMed]
32. Gago, C.M.L.; Artiga-Artigas, M.; Antunes, M.D.C.; Faleiro, M.L.; Miguel, M.G.; Martín-Belloso, O. Effectiveness of nanoemulsions of clove and lemongrass essential oils and their major components against *Escherichia coli* and *Botrytis cinerea*. *J. Food Sci. Technol.* **2019**, *56*, 2721–2736. [CrossRef]
33. Tripathi, A.D.; Sharma, R.; Agarwal, A.; Haleem, D.R. Nanoemulsions based edible coatings with potential food applications. *Int. J. Biobased Plast.* **2021**, *3*, 112–125. [CrossRef]
34. Liu, Q.; Huang, H.; Chen, H.; Lin, J.; Wang, Q. Food-Grade Nanoemulsions: Preparation, Stability and Application in Encapsulation of Bioactive Compounds. *Molecules* **2019**, *24*, 4242. [CrossRef]
35. Zamaniahari, S.; Jamshidi, A.; Moosavy, M.H.; Khatibi, S.A. Preparation and evaluation of *Mentha spicata* L. essential oil nanoemulsion: Physicochemical properties, antibacterial activity against foodborne pathogens and antioxidant properties. *J. Food Meas. Charact.* **2022**, *16*, 3289–3300. [CrossRef]
36. Robledo, N.; Vera, P.; López, L.; Yazdani-Pedram, M.; Tapia, C.; Abugoch, L. Thymol nanoemulsions incorporated in quinoa protein/chitosan edible films; antifungal effect in cherry tomatoes. *Food Chem.* **2018**, *246*, 211–219. [CrossRef] [PubMed]
37. Santamaría Basulto, F.; Sauri Duch, E.; Espadas y Gil, F.; Díaz Plaza, R.; Larqué Saavedra, A.; Santamaría, J.M. Postharvest ripening and maturity indices for maradol papaya. *Interciencia* **2009**, *34*, 583–588.
38. Viacava, G.E.; Ayala-Zavala, J.F.; González-Aguilar, G.A.; Ansorena, M.R. Effect of free and microencapsulated thyme essential oil on quality attributes of minimally processed lettuce. *Postharvest Biol. Technol.* **2018**, *145*, 125–133. [CrossRef]
39. Hagenmaier, R.D.; Baker, R.A. Edible Coatings from Morpholine-Free Wax Microemulsions. *J. Agric. Food Chem.* **1997**, *45*, 349–352. [CrossRef]
40. Miranda, M.; Sun, X.; Ference, C.; Plotto, A.; Bai, J.; Wood, D.; Assis, O.B.G.; Ferreira, M.D.; Baldwin, E. Nano- and Micro-Carnauba Wax Emulsions versus Shellac Protective Coatings on Postharvest Citrus Quality. *J. Am. Soc. Hortic. Sci.* **2021**, *146*, 40–49. [CrossRef]
41. AOAC. *Official Method of Analysis*, 18th ed.; Method 935.14 and 992.24; Association of Officiating Analytical Chemists: Washington, DC, USA, 2005.
42. Martins, D.R.; Barbosa, N.C.; de Resende, E.D. Respiration rate of Golden papaya stored under refrigeration and with different controlled atmospheres. *Sci. Agric.* **2014**, *71*, 369–373. [CrossRef]
43. Miranda, M.; Sun, X.; Marín, A.; dos Santos, L.C.; Plotto, A.; Bai, J.; Benedito Garrido Assis, O.; David Ferreira, M.; Baldwin, E. Nano- and micro-sized carnauba wax emulsions-based coatings incorporated with ginger essential oil and hydroxypropyl methylcellulose on papaya: Preservation of quality and delay of post-harvest fruit decay. *Food Chem.* **2022**, *3*, 100249. [CrossRef] [PubMed]
44. Kamal, I.; Khedr, A.I.M.; Alfaifi, M.Y.; Elbehairi, S.E.I.; Elshaarawy, R.F.M.; Saad, A.S. Chemotherapeutic and chemopreventive potentials of ρ-coumaric acid—Squid chitosan nanogel loaded with Syzygium aromaticum essential oil. *Int. J. Biol. Macromol.* **2021**, *188*, 523–533. [CrossRef]
45. Ferreira, V.R.F.; Brandão, R.M.; Freitas, M.P.; Saczk, A.A.; Felix, F.S.; Silla, J.M.; Teixeira, M.L.; Cardoso, M.D.G. Colorimetric, electroanalytical and theoretical evaluation of the antioxidant activity of *Syzygium aromaticum* L.; *Origanum vulgare* L.; *Mentha spicata* L. and Eremanthus erythropappus M. essential oils, and their major constituents. *New J. Chem.* **2019**, *43*, 7653–7662. [CrossRef]
46. Piras, A.; Porcedda, S.; Falconieri, D.; Maxia, A.; Gonçalves, M.; Cavaleiro, C.; Salgueiro, L. Antifungal activity of essential oil from *Mentha spicata* L. and *Mentha pulegium* L. growing wild in Sardinia island (Italy). *Nat. Prod. Res.* **2021**, *35*, 993–999. [CrossRef] [PubMed]

47. Jayaram, C.S.; Chauhan, N.; Dolma, S.K.; Reddy, S.G.E. Chemical Composition and Insecticidal Activities of Essential Oils against the Pulse Beetle. *Molecules* **2022**, *27*, 568. [CrossRef] [PubMed]
48. Thielmann, J.; Muranyi, P.; Kazman, P. Screening essential oils for their antimicrobial activities against the foodborne pathogenic bacteria Escherichia coli and Staphylococcus aureus. *Heliyon* **2019**, *5*, e01860. [CrossRef]
49. Saadat, S.; Rawtani, D.; Rao, P.K. Antibacterial activity of chitosan film containing Syzygium aromaticum (clove) oil encapsulated halloysite nanotubes against foodborne pathogenic bacterial strains. *Mater. Today Commun.* **2022**, *32*, 104132. [CrossRef]
50. Santolin, L.; Fiametti, K.G.; da Silva Lobo, V.; Wancura, J.H.C.; Oliveira, J.V. Enzymatic Synthesis of Eugenyl Acetate from Essential Oil of Clove Using Lipases in Liquid Formulation as Biocatalyst. *Appl. Biochem. Biotechnol.* **2021**, *193*, 3512–3527. [CrossRef]
51. Xiong, Y.; Kamboj, M.; Ajlouni, S.; Fang, Z. Incorporation of salmon bone gelatine with chitosan, gallic acid and clove oil as edible coating for the cold storage of fresh salmon fillet. *Food Control* **2021**, *125*, 107994. [CrossRef]
52. Adam, K.; Sivropoulou, A.; Kokkini, S.; Lanaras, T.; Arsenakis, M. Antifungal Activities of Origanum vulgare subsp. hirtum, Mentha spicata, Lavandula angustifolia, and Salvia fruticosa Essential Oils against Human Pathogenic Fungi. *J. Agric. Food Chem.* **1998**, *46*, 1739–1745. [CrossRef]
53. Kostik, V.; Gjorgjeska, B.; Petkovska, S.; Mentha, L. essential oils composition and in vitro antifungal activity. *IOSR J. Pharm.* **2015**, *5*, 1–7.
54. Duarte, L.G.R.; Alencar, W.M.P.; Iacuzio, R.; Silva, N.C.C.; Picone, C.S.F. Synthesis, characterization and application of antibacterial lactoferrin nanoparticles. *Curr. Res. Food Sci.* **2022**, *5*, 642–652. [CrossRef]
55. Luksiene, Z.; Buchovec, I. Impact of chlorophyllin-chitosan coating and visible light on the microbial contamination, shelf life, nutritional and visual quality of strawberries. *Innov. Food Sci. Emerg. Technol.* **2019**, *52*, 463–472. [CrossRef]
56. Gao, Y.; Xu, D.; Ren, D.; Zeng, K.; Wu, X. Green synthesis of zinc oxide nanoparticles using Citrus sinensis peel extract and application to strawberry preservation: A comparison study. *LWT* **2020**, *126*, 109297. [CrossRef]
57. Gutiérrez-Pacheco, M.M.; Ortega-Ramírez, L.A.; Silva-Espinoza, B.A.; Cruz-Valenzuela, M.R.; González-Aguilar, G.A.; Lizardi-Mendoza, J.; Miranda, R.; Ayala-Zavala, J.F. Individual and Combined Coatings of Chitosan and Carnauba Wax with Oregano Essential Oil to Avoid Water Loss and Microbial Decay of Fresh Cucumber. *Coatings* **2020**, *10*, 614. [CrossRef]
58. Lin, X.; Zhang, H.; Guo, X.; Qin, Y.; Shen, P.; Peng, Q. A Novel Sodium Alginate-Carnauba Wax Film Containing Calcium Ascorbate: Structural Properties and Preservative Effect on Fresh-Cut Apples. *Molecules* **2023**, *28*, 367. [CrossRef]
59. Baldwin, E.A.; Burns, J.K.; Kazokas, W.; Brecht, J.K.; Hagenmaier, R.D.; Bender, R.J.; Pesis, E. Effect of two edible coatings with different permeability characteristics on mango (*Mangifera indica* L.) ripening during storage. *Postharvest Biol. Technol.* **1999**, *17*, 215–226. [CrossRef]
60. Mendy, T.K.; Misran, A.; Mahmud, T.M.M.; Ismail, S.I. Application of Aloe vera coating delays ripening and extend the shelf life of papaya fruit. *Sci. Hortic.* **2019**, *246*, 769–776. [CrossRef]
61. Sekarina, A.S.; Supriyadi; Munawaroh, H.S.H.; Susanto, E.; Show, P.L.; Ningrum, A. Effects of edible coatings of chitosan—Fish skin gelatine containing black tea extract on quality of minimally processed papaya during refrigerated storage. *Carbohydr. Polym. Technol. Appl.* **2023**, *5*, 100287. [CrossRef]
62. Duarte, L.G.R.; Picone, C.S.F. Antimicrobial activity of lactoferrin-chitosan-gellan nanoparticles and their influence on strawberry preservation. *Food Res. Int.* **2022**, *159*, 111586. [CrossRef] [PubMed]
63. Jiang, Y.; Yu, L.; Hu, Y.; Zhu, Z.; Zhuang, C.; Zhao, Y.; Zhong, Y. Electrostatic spraying of chitosan coating with different deacetylation degree for strawberry preservation. *Int. J. Biol. Macromol.* **2019**, *139*, 1232–1238. [CrossRef] [PubMed]
64. He, Y.; Li, H.; Fei, X.; Peng, L. Carboxymethyl cellulose/cellulose nanocrystals immobilized silver nanoparticles as an effective coating to improve barrier and antibacterial properties of paper for food packaging applications. *Carbohydr. Polym.* **2021**, *252*, 117156. [CrossRef] [PubMed]
65. Tabassum, N.; Khan, M.A. Modified atmosphere packaging of fresh-cut papaya using alginate based edible coating: Quality evaluation and shelf life study. *Sci. Hortic.* **2020**, *259*, 108853. [CrossRef]
66. Vieira, A.C.F.; de Matos Fonseca, J.; Menezes, N.M.C.; Monteiro, A.R.; Valencia, G.A. Active coatings based on hydroxypropyl methylcellulose and silver nanoparticles to extend the papaya (*Carica papaya* L.) shelf life. *Int. J. Biol. Macromol.* **2020**, *164*, 489–498. [CrossRef]
67. Malvano, F.; Corona, O.; Pham, P.L.; Cinquanta, L.; Pollon, M.; Bambina, P.; Farina, V.; Albanese, D. Effect of alginate-based coating charged with hydroxyapatite and quercetin on colour, firmness, sugars and volatile compounds of fresh cut papaya during cold storage. *Eur. Food Res. Technol.* **2022**, *248*, 2833–2842. [CrossRef]
68. Maringgal, B.; Hashim, N.; Amin Tawakkal, I.S.M.; Muda Mohamed, M.T.; Hazwan Hamzah, M.; Ali, M.M.; Abd Razak, M.F.H. Kinetics of quality changes in papayas (*Carica papaya* L.) coated with Malaysian stingless bee honey. *Sci. Hortic.* **2020**, *267*, 109321. [CrossRef]
69. Nazoori, F.; Mollai, S.; Sobhani, F.; Mirdehghan, S.H.; Sahhafi, S.R. Carboxymethyl cellulose and carnauba wax treatments kept the pomegranate fruit (*Punica granatum* L.) quality during cold storage via improving enzymatic defense system and bioactive compounds. *Sci. Hortic.* **2023**, *309*, 111645. [CrossRef]
70. Tan, G.H.; Ali, A.; Siddiqui, Y. Current strategies, perspectives and challenges in management and control of postharvest diseases of papaya. *Sci. Hortic.* **2022**, *301*, 111139. [CrossRef]

71. Raybaudi-Massilia, R.M.; Mosqueda-Melgar, J.; Martín-Belloso, O. Edible alginate-based coating as carrier of antimicrobials to improve shelf-life and safety of fresh-cut melon. *Int. J. Food Microbiol.* **2008**, *121*, 313–327. [CrossRef]
72. Chen, H.; Sun, Z.; Yang, H. Effect of carnauba wax-based coating containing glycerol monolaurate on the quality maintenance and shelf-life of Indian jujube (*Zizyphus mauritiana Lamk.*) fruit during storage. *Sci. Hortic.* **2019**, *244*, 157–164. [CrossRef]
73. Liu, Y.; Tang, M.; Liu, M.; Su, D.; Chen, J.; Gao, Y.; Bouzayen, M.; Li, Z. The Molecular Regulation of Ethylene in Fruit Ripening. *Small Methods* **2020**, *4*, 1900485. [CrossRef]
74. Choi, W.S.; Singh, S.; Lee, Y.S. Characterization of edible film containing essential oils in hydroxypropyl methylcellulose and its effect on quality attributes of 'Formosa' plum (*Prunus salicina* L.). *LWT* **2016**, *70*, 213–222. [CrossRef]
75. Özden, Ç.; Bayindirli, L. Effects of combinational use of controlled atmosphere, cold storage and edible coating applications on shelf life and quality attributes of green peppers. *European Food Res. Technol.* **2002**, *214*, 320–326. [CrossRef]
76. Boshkova, N.; Stambolova, I.; Stoyanova, D.; Simeonova, S.; Grozev, N.; Avdeev, G.; Shipochka, M.; Dimitrov, O.; Bachvarov, V.; Peshova, M.; et al. Protective Characteristics of TiO_2 Sol-Gel Layer Deposited on Zn-Ni or Zn-Co Substrates. *Coatings* **2023**, *13*, 295. [CrossRef]
77. Xue, F.; Gu, Y.; Wang, Y.; Li, C.; Adhikari, B. Encapsulation of essential oil in emulsion based edible films prepared by soy protein isolate-gum acacia conjugates. *Food Hydrocoll.* **2019**, *96*, 178–189. [CrossRef]
78. Ziani, K.; Fang, Y.; McClements, D.J. Fabrication and stability of colloidal delivery systems for flavor oils: Effect of composition and storage conditions. *Food Res. Int.* **2012**, *46*, 209–216. [CrossRef]
79. Antonioli, G.; Fontanella, G.; Echeverrigaray, S.; Longaray Delamare, A.P.; Fernandes Pauletti, G.; Barcellos, T. Poly(lactic acid) nanocapsules containing lemongrass essential oil for postharvest decay control: In vitro and in vivo evaluation against phytopathogenic fungi. *Food Chem.* **2020**, *326*, 126997. [CrossRef]
80. Vellido-Perez, J.A.; Ochando-Pulido, J.M.; Brito-de la Fuente, E.; Martinez-Ferez, A. Novel emulsions–based technological approaches for the protection of omega–3 polyunsaturated fatty acids against oxidation processes—A comprehensive review. *Food Struct.* **2021**, *27*, 100175. [CrossRef]

coatings

MDPI

Communication

Effects of Commercial Natural Compounds on Postharvest Decay of Strawberry Fruit

Razieh Rajestary †, Lucia Landi † and Gianfranco Romanazzi *

Department of Agricultural, Food and Environmental Sciences, Marche Polytechnic University, Via Brecce Bianche, I-60131 Ancona, Italy; sabasol2002@yahoo.com (R.R.); l.landi@univpm.it (L.L.)
* Correspondence: g.romanazzi@univpm.it
† These authors contributed equally to this work.

Abstract: Gray mold and Rhizopus rot, which is caused by *Botrytis cinerea* and *Rhizopus stolonifer*, respectively, are the most destructive forms of postharvest decay of the strawberry fruit. In this work, we tested the effectiveness of the control on the postharvest decay of the strawberry fruit (*Fragaria* × *ananassa* Duch cv. 'Monterey') following postharvest applications of six commercial natural compounds: chitosan-based coating compound (1% of 'ChitP', 'ChitS', 'ChitK', 'ChitO'), commercial essential oil (EOs) products based on grapefruit seed extract (0.5% of 'GraFr'), sweet orange (0.5% of 'SwOr'), a product that included eugenol, geraniol, and thymol EO, (0.4% of 'EuGeTh'), an organic compound as humic acid (0.5% w/v of 'HuAc'), and, lastly, methyl jasmonate plant growth regulator (1% v/v 'MeJA'). Strawberries were dipped in solution for 30 s and incubated at room temperature (20 ± 0.5 °C) or at cold storage conditions (4 ± 0.5 °C) following 4 days of shelf life at 20 °C. The treatments with 'ChitP', 'ChitS', and 'ChitO' provided ~30%–40% reduction of gray mold in cold storage conditions, while the 'MeJA', 'SwOr', and 'GraFr' with high activities of volatile substances were more effective at controlling gray mold at room temperature. 'HuAc', 'ChitK', and 'ChitO' were more effective at controlling Rhizopus rot in both cold storage (~50%) and room temperature conditions.

Keywords: basic substances; *Botrytis cinerea*; *Rhizopus stolonifer*; strawberry

Citation: Rajestary, R.; Landi, L.; Romanazzi, G. Effects of Commercial Natural Compounds on Postharvest Decay of Strawberry Fruit. *Coatings* **2023**, *13*, 1515. https://doi.org/10.3390/coatings13091515

Academic Editor: Swarna Jaiswal

Received: 25 April 2023
Revised: 16 August 2023
Accepted: 24 August 2023
Published: 28 August 2023

1. Introduction

The strawberry fruit (*Fragaria* × *ananassa* Duch.) is highly appreciated by consumers for its unique taste and flavor as well as its health benefits and exceptional nutritional value [1,2]. Indeed, strawberries are rich in bioactive compounds, such as natural antioxidants, polyphenols, anthocyanins, vitamins, and amino acids [3–7]. However, strawberries are particularly perishable, especially during postharvest storage, and they are susceptible to both mechanical damage and fungal disease, which limits their commercialization and consumption [8]. Gray mold and Rhizopus soft rot caused by *Botrytis cinerea* (Pers.) and *Rhizopus stolonifer* (Ehrenb.), respectively, are the main pathogens of the postharvest decay of the strawberry [9,10]. A primary infection of gray mold could occur at bloom time and remain quiescent in the field [10,11]. *B. cinerea* produces large numbers of spores, and it was able to survive in a dormant state in a variety of environmental conditions. [12]. Therefore, it is not surprising that *B. cinerea* ranked second in the top 10 fungal plant pathogens list based on scientific and economic importance [13]. *R. stolonifer* is a common wound pathogen of a very wide range of fruits and vegetables, causing a rapidly spreading watery soft rot. Rhizopus rot can spread at temperatures greater than 4–6 °C. Both gray mold and Rhizopus soft rot spread quickly to other fruit, and this phenomenon is known as nesting [10,14].

Despite the effectiveness of the synthetic fungicides in the management of strawberry fruit disease, natural eco-friendly alternative compounds are desirable, and they have attracted the attention of scientists, who aim to provide growers, consumers, and the whole community with information on the strategies that are effective and, at the same time, safer for consumers.

In recent years, the antimicrobial activity of a large number of compounds similar to plant and animal extracts, such as gums, resins, etc., have been tested against both pre-and postharvest pathogens [15,16]. These compounds were non-toxic for human health and the environment, had no negative effects on the quality of the fruits, and might complement or even improve current productive practices. Natural compounds are characterized by antimicrobial activities against the main postharvest pathogens and/or are resistance inducers that activate plant defenses in order to simulate the presence of a pathogen.

Among the natural compounds, chitosan has received much attention for its application in agriculture and in the food industry. Chitosan is a natural biopolymer, derived from chitin of both marine crustaceans [17] and the cell wall of many pathogenic fungi [18–20]. This compound has been reported to stimulate plant defenses and prevent disease development [15,21]. A number of promising approaches for the postharvest application of different types of chitosan formulation have been suggested [22–24], and the effectiveness of the combination of chitosan with essential oils (EOs) has also been observed [25–27].

Essential oils from aromatic plants have been gaining interest, and their effectiveness at controlling the postharvest decay of fruit has been documented [28–32]. Other compounds, such as humic acid [33,34], an organic compound known as a promoter in sustaining plant growth [35], have been reported to have efficacy in the control of several plant diseases, inducing host resistance and direct antimicrobial activity [36]. In the same way, methyl jasmonate (MeJA) [37] is an endogenous plant growth substance that can modulate many physiological processes, including responses to environmental stress [38].

The objective of this study was to verify the effectiveness of a list of promising commercial compounds (listed in Table 1) based on chitosan, EOs, organic compounds, and plant growth regulator on the control of the postharvest decay of strawberries kept at either room temperature or cold stored and then exposed to shelf life.

Table 1. Commercial names and sources of the formulations containing the active ingredients used in the postharvest treatments of strawberries.

Name	Formulation Commercial Name	Source (Country)	Active Ingredient	Application Dose (*v/v*); (*w/v*) *
'ChitP'	Chito Plant powder	ChiPro GmbH; (Bremen, Germany)	Chitosan	1% *
'ChitS'	Chito Plant Solution	ChiPro GmbH; (Bremen, Germany)	Chitosan	1%
'ChitO'	OII-YS	Venture Chemicals, Inc.; (Lafayette, LA, USA)	Chitosan	1%
'ChitK'	Kaitosol	Advanced Green Nanotechnologies Sdn Bhd; (Cambridge, UK)	Chitosan	1%
'GraFr'	DF-100 Forte	Agritalia, (Rovigo, Italy)	Grapefruit seed extract	0.5%
'SwOr'	Prev-Am plus	Nufram, (Milano, Italy)	Sweet orange extract	0.5%
'EuGeTh'	3Logy	Sipcam, (Milano, Italy)	Eugenol, geraniol, and thymol extracts	0.4%
'HuAc'	Humic acid	Sigma-Aldrich, (Saint Louis, MO, USA)	Humic acid sodium salt	0.5% *
'MeJA'	Methyl jasmonate	Sigma-Aldrich, (Saint Louis, MO, USA)	Methyl jasmonate	1%

* = weight by volume solution (*w*/*v*).

2. Materials and Methods

2.1. Fruit Material

Commercial strawberries (*Fragaria* × *ananassa* Duch, cv 'Monterey') were collected from an orchard located in Montalto (AP) in the Marche region in central-eastern Italy. The strawberries were harvested at the mature stage, and were selected for the absence of defects, uniformity in size, and the degree of ripening (2/3 red on the surface) [39]. They were used for the experiments on the day that they were harvested [14].

2.2. Preparation of Natural Compounds Solution

A list of chitosan-based commercial compounds available on the market together with other formulation alternatives to synthetic fungicides that could have an effect on the postharvest decay of strawberries were included in the investigation. The compounds used for the postharvest treatments are summarized in Table 1. All of the compounds were dissolved in Tween 80, 20 μL/L (Sigma Chemical Co., St. Louis, MO, USA) water solution for 1 h.

2.3. Postharvest Treatments

The strawberries were immersed in the solutions ready to be tested according to Feliziani et al. [15]. In detail, the strawberries were soaked for 30 s inside each solution, air dried for 3 to 4 h, and then arranged in small plastic boxes. They were incubated in two different conditions: room temperature (20 ± 0.5 °C) and cold temperature (4 ± 0.5 °C) for 7 days, 95%–98% RH, and they were then exposed to 4 days of shelf life at 20 °C, 95%–98% RH. Each treatment consisted of 66 fruits (6 fruits in 11 plastic boxes). Three replications were performed for each treatment. The infections that subsequently developed resulted in naturally occurring inoculum for the following treatments: (i.) natural compound solution (treated strawberry fruit), and (ii.) sterile distilled water (untreated strawberry fruit).

2.4. Data Recording

During storage, data were recorded based on the percentage of the incidence of decay on the strawberries. Disease severity was also measured according to an empirical scale with five degrees: 0, healthy fruit; 1, 1%–20% fruit surface infected; 2, 21%–40% fruit surface infected; 3, 41%–60% fruit surface infected; 4, 61%–80% fruit surface infected; and 5, more than 81% of the strawberry surface infected and showing sporulation [15]. The empirical scale allowed the calculation of the McKinney index, which was expressed as the weighted average of the disease as a percentage of the maximum possible level [40,41]. This parameter also included information on both disease incidence and disease severity.

2.5. Statistical Analysis

Statistical analysis was performed based on the Fisher test. Differences among the means of the values were analyzed by one-way analysis of variance (ANOVA). Difference was considered as statistically significant at $p < 0.05$. Moreover, the treatments were subjected to rank analysis that allowed us to combine heterogeneous data (Excel 2007) [42,43].

3. Results

Decay Evaluation

The postharvest treatments with commercial compounds generally reduced the development of the decay of the strawberries after 4 days of shelf life at both room temperature (20 ± 1 °C) and cold temperature (4 ± 1 °C), which was mainly gray mold followed by Rhizopus rot. However, the more significant decrease in both disease incidence and severity was observed in the cold temperature condition (data not shown). The McKinney index of decay was significantly decreased compared to the control: the compounds based on chitosan, 'ChitP', 'ChitS', 'ChiK', and 'ChiO', had decreases of 35.36%, 26.82%, 24.39, and 45.12%, respectively, whilst the compounds based on EOs, 'GraFr', and 'EuGeTh', had decreases of 28.65% and 29.26%, respectively, and, finally, those with 'MeJA' and 'HuAc' had decreases of 31.7% and 32.92%, respectively (Figure 1).

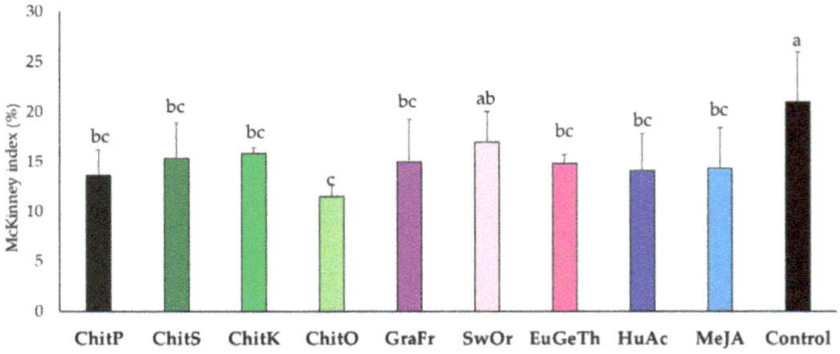

Figure 1. McKinney's index of gray mold of the 'Monterey' strawberry fruit. Strawberries were treated after harvest, stored for 7 days at 4 ± 0.5 °C, and then exposed to 4 days of shelf life at 20 ± 1 °C and 95% to 98% relative humidity. Values with different small letters are different at $p < 0.05$. Note: 'ChitP' = Chito Plant powder; 'ChitS' = Chito Plant solution; 'ChitK' = KaitoSol; 'ChitO' = OII-YS; 'GraFr' = DF-100 Forte; 'SwOr' = Prev-Am plus; 'EuGeTh' = 3Logy; 'HuAc' = Humic acid; 'MeJA' = methyl jasmonate.

The treatment with the 'SwOr' decreased the McKinney index of decay by 19.26%, although it did not show a significant reduction compared to the control. A more direct analysis of the degree of comparative effectiveness for the reduction of disease incidence was obtained through the application of rank analysis. At both room temperature and cold storage conditions, the untreated fruits had the highest sum of ranks, namely, 8.5 and 8.6, respectively, and, therefore, all of the treatments were more effective compared to the control (Figure 2). However, some differences occurred among the treatments at different storage temperatures. The commercial compounds 'ChitP', 'ChitS', and 'ChitO' were more effective at controlling postharvest disease in strawberries in cold storage conditions (sum of ranks 2.2, 3.1, and 5.2, respectively) compared to room temperature storage (sum of ranks 3.9, 4.1, and 7.5, respectively) (Figure 2). In contrast, the 'MeJA', 'SwOr', and 'GraFr' were more effective at controlling postharvest disease in strawberries at room temperature conditions (4.8, 2.5, and 2, respectively) compared to cold storage ones (8.4, 5.8, and 6.5, respectively). The other compounds tested showed efficiency at controlling postharvest rot in strawberries that was similar to the two storage conditions that we tested (Figure 2).

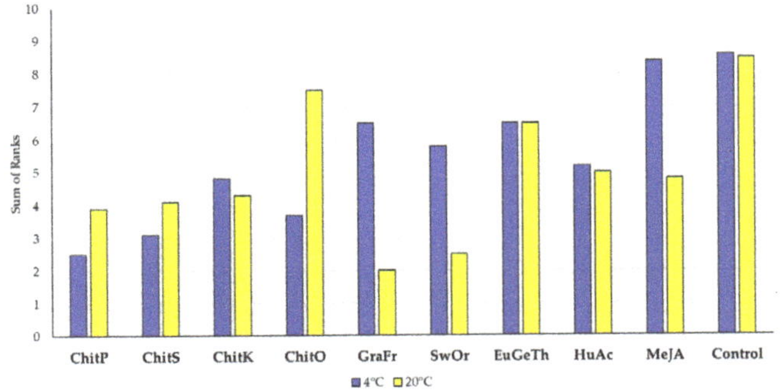

Figure 2. The effect of postharvest treatment with natural compounds on the reduction of gray mold on strawberries according to rank analysis. The fruit was kept at 4 °C and 20 ± 1 °C, 95%–98% RH. Note: 'ChitP' = 'Chito Plant powder; 'ChitS' = Chito Plant solution; 'ChitK' = KaitoSol; 'ChitO' = OII-YS; 'GraFr' = DF-100 Forte; 'SwOr' = Prev-Am plus; 'EuGeTh' = 3Logy; 'HuAc' = Humic acid; 'MeJA' = Methyl jasmonate.

Based on cumulative incidence, the effectiveness of different natural compounds against *R. stolonifer* on strawberries was measured. The Rhizopus rot cumulative incidence for all of the successful edible coating was about 1%–2% lower than the control at both temperatures. However, 'HuAc', 'ChitK', and 'ChitO' were the most successful compounds at reducing Rhizopus rot at cold storage conditions, and the cumulative incidence for these compounds was less than half of the control (data not shown).

4. Discussion

The present study shows that compounds from natural sources, such as chitosan different emulsions, commercial EOs, and organic plant growth regulator compounds with promising properties, can reduce the development of postharvest rots in strawberry fruits. All of the tested compounds significantly reduced decay on cold-stored strawberry fruits, and the best results were observed using the chitosan compounds. On strawberries kept at room temperature, the rank analysis showed that all of the tested compounds were effective at decay control compared to the control. The commercial products tested decreased the development of gray mold on strawberries, prolonging the shelf life of the fruit. Based on rank analysis, the effectiveness of the tested compounds was different according to the storage temperature of the strawberries: 'GraFr' and 'SwOr' provided the highest reduction of gray mold (76.4% and 70.5%, respectively, compared to the control) on strawberries kept at room temperature, while the compounds based on chitosan, 'ChitP' and 'ChitP', showed the best performance on cold-stored fruit (76.4% and 63.9%, respectively, compared to the control). The higher effectiveness of 'GraFr' and 'SwOr' at room temperature can be ascribed to their high activity of volatile composition. A similar result was also observed for the 'MeJa', a volatile compound that is an important cellular regulator, and which is able to reduce the gray mold and brown rot, thereby extending the shelf life of fruits [43,44]. Room temperature crucially influences the stability of EOs in several aspects. On these lipophilic compounds, which are highly volatile and plant secondary metabolites, the chemical reactions generally accelerate with increasing heat [45,46]. Consequently, the application method can affect the efficacy of postharvest treatments of EOs [29], as has been observed for the EO of oregano, red thyme, peppermint, and lemongrass incorporated in chitosan coatings on strawberry fruits [47]. Strong antifungal activity from the above EOs could be attributed to their components [48,49]. 'SwOr' and 'GraFr' consisted of sweet orange essential oils and grapefruit seed extract, respectively. The composition of 'EuGeTh' included eugenol, geraniol, and thymol, which are very well known for their bioactivity against fungal pathogens [50–52]. The activity of 'EuGeTh' as a biocontrol agent for grape vineyards against gray mold has also been observed [32]. In our work, we did not detect the same effectiveness on the postharvest strawberry treatment. Among the EO-based compounds, 'EuGeTh' was the least effective in the control of the storage decay of strawberries. Concerning the compounds based on chitosan, the refrigerated storage was effective in maintaining the postharvest quality of strawberries. The effectiveness of chitosan in disease control showed triple activity associated with antimicrobial activity, host defense activation, and film formation on the treated surface [19,53,54]. Previous works estimated that chitosan is one of the most effective alternative compounds to control the disease and prolong shelf life at cold storage conditions. It is known that chitosan coatings delay changes in weight loss, soluble solids, and total sugars, and reduce the ethylene production; these actions could be improved at low temperature conditions, leading to a lower disease incidence of fungal pathogen [53,55]. Chitosan is one of the most common resistance inducers available on the market, and elicitation of host defenses allows postharvest decay to be managed, limiting the application of synthetic pesticides and increasing the production of nutraceutical compounds [56].

5. Conclusions

The tested natural compounds were effective at both cold storage and room temperature at containing the postharvest decay of strawberries, and they had a variable action according to the storage conditions. For cold-stored strawberries, all of the tested compounds, with the exception of 'SwOr', were effective at reducing gray mold infections. Overall, chitosan formulations, including 'ChitP', 'ChitO', and 'ChitS', were the most effective compounds for controlling *B. cinerea*, while the compounds based on EOs, 'SwOr' and 'GraFr', showed the highest effectiveness at room temperature. Our work emphasizes that storage temperature and the formulation of compounds are both factors that influence the effectiveness of the compounds. However, our work was run with the immersion of the strawberry fruit, and to progress to practical application, field experiments will be necessary.

Author Contributions: R.R.: formal analysis, data curation, writing; L.L.: data curation, writing, review and editing; G.R.: conceptualization, review and Editing All authors have read and agreed to the published version of the manuscript.

Funding: This work was conducted within PRIMA StopMedWaste 'Innovative Sustainable technologies TO extend the shelf-life of Perishable MEDiterranean fresh fruit, vegetables and aromatic plants and to reduce WASTE', which are funded by PRIMA, a program supported by the European Union, and Euphresco BasicS 'Basic substances as an environmentally friendly alternative to synthetic pesticides for plant protection' projects.

Institutional Review Board Statement: Not applicable.

Informed Consent Statement: Not applicable.

Data Availability Statement: Not applicable.

Conflicts of Interest: The authors declare no conflict of interest.

References

1. Mikulic-Petkovsek, M.; Schmitzer, V.; Slatnar, A.; Weber, N.; Veberic, R.; Stampar, F.; Munda, A.; Koron, D. Alteration of the content of primary and secondary metabolites in strawberry fruit by *Colletotrichum nymphaeae* Infection. *J. Agric. Food Chem.* **2013**, *61*, 5987–5995. [CrossRef]
2. Van de Velde, F.; Tarola, A.M.; Guemes, D.; Pirovani, M.E. Bioactive compounds and antioxidant capacity of Camarosa and Selva Strawberries (*Fragaria ananassa* Duch.). *Foods* **2013**, *2*, 120–131. [CrossRef]
3. Afrin, S.; Gasparrini, M.; Forbes-Hernandez, T.Y.; Reboredo-Rodriguez, P.; Mezzetti, B.; Varela-López, A.; Giampieri, F.; Battino, M. Promising health benefits of the strawberry: A focus on clinical studies. *J. Agric. Food Chem.* **2016**, *64*, 4435–4449. [CrossRef]
4. Afrin, S.; Giampieri, F.; Cianciosi, D.; Alvarez-Suarez, J.M.; Bullon, B.; Amici, A.; Quiles, J.L.; Forbes-Hernández, T.Y.; Battino, M. Strawberry tree honey in combination with 5-fluorouracil enhances chemosensitivity in human colon adenocarcinoma cells. *Food Chem. Toxicol.* **2021**, *156*, 112484. [CrossRef]
5. Ezzat, M.I.; Issa, M.Y.; Sallam, I.E.; Zaafar, D.; Khalil, H.M.A.; Mousa, M.R.; Sabry, D.; Gawish, A.Y.; Elghandour, A.H.; Mohsen, E. Impact of different processing methods on the phenolics and neuroprotective activity of *Fragaria ananassa* Duch. extracts in a d-galactose and aluminum chloride-induced rat model of aging. *Food Funct.* **2022**, *13*, 7794–7812. [CrossRef]
6. Li, R.; Tao, M.; Xu, T.; Pan, S.; Xu, X.; Wu, T. Small berries as health-promoting ingredients: A review on anti-aging effects and mechanisms in *Caenorhabditis elegans*. *Food Funct.* **2022**, *13*, 478–500. [CrossRef]
7. Sánchez-Velázquez, O.A.; Mulero, M.; Cuevas-Rodríguez, E.O.; Mondor, M.; Arcand, Y.; Hernández-Álvarez, A.J. In vitro gastrointestinal digestion impact on stability, bioaccessibility and antioxidant activity of polyphenols from wild and commercial blackberries (*Rubus* spp.). *Food Funct.* **2021**, *12*, 7358–7378. [CrossRef]
8. Ugolini, L.; Martini, C.; Lazzeri, L.; Avinoa, L.D.; Mari, M. Control of postharvest grey mould (*Botrytis cinerea* Per. Fr.) on strawberries by glucosinolate-derivedallyl-isothiocyanate treatment. *Postharvest Biol. Technol.* **2014**, *90*, 34–39. [CrossRef]
9. Maas, J.L. *Compendium of Strawberry Disease*; APS Press: St. Paul, MN, USA, 1998; 128p.
10. Feliziani, E.; Romanazzi, G. Postharvest decay of strawberry fruit: Etiology, epidemiology, and disease management. *J. Berry Res.* **2016**, *6*, 47–63. [CrossRef]
11. Powelson, E.L. Initiation of strawberry fruit rot caused by *Botrytis cinerea*. *Phytopathology* **1960**, *50*, 491–494.
12. Hahn, M. The rising threat of fungicide resistance in plant pathogenic fungi: *Botrytis* as a case study. *J. Chem. Biol.* **2014**, *7*, 133–141. [CrossRef]
13. Dean, R.; Van Kan, J.A.; Pretorius, Z.A.; Hammond-Kosack, K.E.; Di Pietro, A.; Spanu, P.D.; Rudd, J.J.; Dickman, M.; Kahmann, R.; Ellis, J.; et al. The top 10 fungal pathogens in molecular plant pathology. *Mol. Plant Pathol.* **2012**, *13*, 414–430. [CrossRef] [PubMed]
14. Ugolini, L.; Pagnotta, E.; Matteo, R.; Malaguti, L.; Di Francesco, A.; Lazzeri, L. Brassica meal-derived allyl-isothiocyanate postharvest application: Influence on strawberry nutraceutical and biochemical parameters. *J. Sci. Food Agric.* **2019**, *99*, 4235–4241. [CrossRef] [PubMed]

15. Feliziani, E.; Landi, L.; Romanazzi, G. Preharvest treatments with chitosan and other alternatives to conventional fungicides to control postharvest decay of strawberry. *Carbohydr. Polym.* **2015**, *132*, 111–117. [CrossRef]
16. Lelario, F.; Scrano, L.; De Franchi, S.; Bonomo, M.G.; Salzano, G.; Milan, S.; Milella, L.; Bufo, S.A. Identification and antimicrobial activity of most representative secondary metabolites from different plant species. *Chem. Biol. Technol. Agric.* **2018**, *5*, 13. [CrossRef]
17. Dima, C.; Patrascu, L.; Cantaragiu, A.; Alexe, P.; Dima, S. The kinetics of the swelling process and the release mechanisms of *Coriandrum sativum* L. essential oil from chitosan/alginate/inulin microcapsules. *Food Chem.* **2016**, *195*, 39–48. [CrossRef]
18. Benhabiles, M.S.; Salah, R.; Lounici, H.; Drouiche, N.; Goosen, M.F.A.; Mameri, N. Antibacterial activity of chitin, chitosan and its oligomers prepared from shrimp shell waste. *Food Hydrocoll.* **2012**, *29*, 48–56. [CrossRef]
19. Romanazzi, G.; Feliziani, E.; Bautista-Banos, S.; Sivakumar, D. Shelf life extension of fresh fruit and vegetables by chitosan treatment. *Crit. Rev. Food Sci. Nutr.* **2017**, *57*, 579–601. [CrossRef]
20. Rajestary, R.; Landi, L.; Romanazzi, G. Chitosan and postharvest decay of fresh fruit: Meta-analysis of disease control and antimicrobial and eliciting activities. *Compr. Rev. Food Sci. Food Saf.* **2021**, *20*, 563–582. [CrossRef]
21. Landi, L.; De Miccolis Angelini, R.M.; Pollastro, S.; Feliziani, E.; Faretra, F.; Romanazzi, G. Global transcriptome analysis and identification of differentially expressed genes in strawberry after preharvest application of benzothiadiazole and chitosan. *Front. Plant Sci.* **2017**, *8*, 235. [CrossRef]
22. Gramisci, B.R.; Lutz, M.C.; Lopes, C.A.; Sangorrína, M.P. Enhancing the efficacy of yeast biocontrol agents against postharvest pathogens through nutrient profiling and the use of other additives. *Biol. Control* **2018**, *121*, 151–158. [CrossRef]
23. Liu, C.; Yang, L.; Qiao, L.; Liu, C.; Zhang, M.; Jian, X. An injectable and self-healing novel chitosan hydrogel with low adamantane substitution degree. *Polym. Int.* **2019**, *68*, 1102–1112. [CrossRef]
24. Basumatary, I.B.; Mukherjee, A.; Katiyar, V.; Kumar, S.; Dutta, J. Chitosan-based antimicrobial coating for improving postharvest shelf life of pineapple. *Coatings* **2021**, *11*, 1366. [CrossRef]
25. Barreto, T.A.; Andrade, S.C.A.; Maciel, J.F.; Arcanjo, N.M.O.; Madruga, M.S.; Meireles, B.; Cordeiro, Â.M.T.; Souza, E.L.; Magnani, M.A. Chitosan coating containing essential oil from *Origanum vulgare* L. to control postharvest mold infections and keep the quality of cherry tomato Fruit. *Front. Microbiol.* **2016**, *7*, 1724. [CrossRef] [PubMed]
26. Landi, L.; Peralta-Ruiz, Y.; Chaves-López, C.; Romanazzi, G. Chitosan coating enriched with *Ruta graveolens* L. essential oil reduces postharvest anthracnose of papaya (*Carica papaya* L.) and modulates defense-related gene expression. *Front. Plant Sci.* **2021**, *12*, 765806. [CrossRef] [PubMed]
27. Salgado-Cruz, M.d.l.P.; Salgado-Cruz, J.; García-Hernández, A.B.; Calderón-Domínguez, G.; Gómez-Viquez, H.; Oliver-Espinoza, R.; Fernández-Martínez, M.C.; Yáñez-Fernández, J. Chitosan as a coating for biocontrol in postharvest products: A bibliometric review. *Membranes* **2021**, *11*, 421. [CrossRef]
28. De Corato, U.; Maccioni, O.; Trupo, M.; Di Sanzo, G. Use of essential oil of *Laurus nobilis* obtained by means of a supercritical carbon dioxide technique against postharvest spoilage fungi. *Crop Prot.* **2010**, *29*, 142–147. [CrossRef]
29. Lopez-Reyes, J.G.; Spadaro, D.; Prelle, A.; Garibaldi, A.; Gullino, M.L. Efficacy of plant essential oils on postharvest control of rots caused by fungi on different stone fruits in vivo. *J. Food Prot.* **2013**, *76*, 631–639. [CrossRef]
30. Sivakumar, D.; Bautista-Baños, S. A review on the use of essential oils for postharvest decay control and maintenance of fruit quality during storage. *Crop Prot.* **2014**, *64*, 27–37. [CrossRef]
31. Mari, M.; Bautista Baños, S.; Sivakumar, D. Decay control in the postharvest system: Role of microbial and plant volatile organic compounds. *Postharvest Biol. Technol.* **2016**, *122*, 70–81. [CrossRef]
32. Rotolo, C.; De Miccolis Angelini, R.M.; Dongiovanni, C.; Pollastro, S.; Fumarola, G.; Carolo, M.D.C.; Perrelli, D.; Natale, P.; Faretra, F. Use of biocontrol agents and botanicals in integrated managment of *Botrytis cinerea* in table grape vineyards. *Pest Manag. Sci.* **2017**, *74*, 715–725. [CrossRef] [PubMed]
33. Gholami, H.; Saharkhiz, M.J.; Fard, F.R.; Ghani, A.; Nadaf, F. Humic acid and vermicompost increased bioactive components, antioxidant activity and herb yield of chicory (*Cichorium intybus* L.). *Biocatal. Agric. Biotechnol.* **2018**, *14*, 286–292. [CrossRef]
34. Wang, L.; Dou, G.; Guo, H.; Zhang, Q.; Qin, X.; Yu, W.; Jiang, C.; Xiao, H. *Hanseniaspora uvarum* prolongs shelf life of strawberry via volatile production. *Nutr. Food Sci.* **2019**, *7*, 2625–2636. [CrossRef]
35. Suh, H.Y.; Yoo, K.S.; Sang, G.S. Effect of foliar application of fulvic acid on plant growth and fruit quality of tomato (*Lycopersicon esculentum* L.). *Hortic. Environ. Biotechnol.* **2014**, *55*, 455–461. [CrossRef]
36. Xu, D.; Deng, Y.; Xi, P.; Yu, G.; Wang, Q.; Zeng, Q.; Jiang, Z.; Gao, L. Fulvic acid-induced disease resistance to *Botrytis cinerea* in table grapes maybe mediated by regulating phenylpropanoid metabolism. *Food Chem.* **2019**, *286*, 226–233. [CrossRef]
37. Glowacz, M.; Roets, N.; Sivakumar, D. Control of anthracnose disease via increased activity of defence related enzymes in 'Hass' avocado fruit treated with methyl jasmonate and methyl salicylate. *Food Chem.* **2017**, *234*, 163–167. [CrossRef]
38. Greelman, R.A.; John, E.; Mullet, J.E. Biosynthesis and action of jasmonates in plants. *Annu. Rev. Plant Physiol. Plant Mol. Biol.* **1997**, *48*, 355–381. [CrossRef] [PubMed]
39. Rosati, P.; Cantoni, L. Fragola: Tecnologia e ricerche avanzate nel mondo. *Frutticoltura* **1993**, *55*, 9–27.
40. McKinney, H.H. Influence of soil temperature and moisture on infection of wheat seedlings by *Helmintosporium sativum*. *J. Agric. Res.* **1923**, *26*, 195–218.
41. Conover, W.J.; Iman, L.R. Rank transformations as a bridge between parametric and nonparametric statistics. *Am. Stat.* **1981**, *35*, 124–129.

42. Romanazzi, G.; Mlikota Gabler, F.; Margosan, D.A.; Mackey, B.E.; Smilanick, J.L. Effect of chitosan dissolved in different acids on its ability to control postharvest gray mold of table grapes. *Phytopathology* **2009**, *99*, 1028–1036. [CrossRef]
43. Wang, X.; Kou, C.; Wu, G.; Fan, T. Li Methyl jasmonate induces the resistance of postharvest blueberry to gray mold caused by *Botrytis cinerea*. *J. Sci. Food Agric.* **2020**, *100*, 4272–4281. [CrossRef]
44. Ezzat, A.; Szabó, S.; Szabó, Z.; Hegedűs, A.; Berényi, D.; Holb, I.J. Temporal patterns and inter-correlations among physical and antioxidant attributes and enzyme activities of apricot fruit inoculated with *Monilinia laxa* under salicylic acid and methyl jasmonate treatments under shelf-life conditions. *J. Fungi* **2021**, *7*, 341. [CrossRef] [PubMed]
45. Peng, Y.; Yin, L.; Li, Y. Combined effects of lemon essential oil and surfactants on physical and structural properties of chitosan films. *Int. J. Food Sci.* **2012**, *48*, 44–50. [CrossRef]
46. Hosseini, S.F.; Amraie, M.; Salehi, M.; Mohseni, M.; Aloui, H. Effect of chitosan-based coatings enriched with savory and/or tarragon essential oils on postharvest maintenance of kumquat (*Fortunella* sp.) fruit. *Nutr. Food Sci.* **2019**, *7*, 155–162. [CrossRef]
47. Vu, K.D.; Hollingsworth, R.G.; Leroux, E.; Salmieri, S.; Lacroix, M. Development of edible bioactive coating based on modified chitosan for increasing the shelf life of strawberries. *Int. Food Res. J.* **2011**, *44*, 198–203. [CrossRef]
48. Císarová, M.; Tančinová, D.; Medo, J. Antifungal activity of lemon, eucalyptus, thyme, oregano, sage and lavender essential oils against *Aspergillus niger* and *Aspergillus tubingensis* isolated from grapes. *Potr. Sci. J. Food Ind.* **2016**, *10*, 83–88. [CrossRef]
49. Mahato, N.; Sharma, K.; Koteswararao, R.; Sinha, M.; Baral, E.; Hwan Cho, M. Citrus essential oils: Extraction, authentication and application in food preservation. *Crit. Rev. Food Sci. Nutr.* **2019**, *59*, 611–625. [CrossRef]
50. Zhou, D.; Wang, Z.; Li, M.; Xing, M.; Xian, T.; Tu, K. Carvacrol and eugenol effectively inhibit *Rhizopus stolonifer* and control postharvest soft rot decay in peaches. *J. Appl. Microbiol.* **2017**, *124*, 166–178. [CrossRef] [PubMed]
51. Zhou, L.; Zhang, Z.; Wei, M.; Xie, Y.; He, S.; Shi, H.; Lin, Z. Evaluation of the antifungal activity of individual and combined monoterpenes against *Rhizopus stolonifer* and *Absidia coerulea*. *Environ. Sci. Pollut. Res.* **2019**, *26*, 7804–7809. [CrossRef]
52. Oliveira, J.; Parisi, M.C.M.; Baggio, J.S.; Silva, P.P.M.; Paviani, B.; Spoto, M.H.F.; Glori, E.M. Control of *Rhizopus stolonifer* in strawberries by the combination of essential oil with carboxymethylcellulose. *Int. J. Food Microbiol.* **2019**, *292*, 150–158. [CrossRef]
53. Romanazzi, G.; Feliziani, E.; Sivakumar, D. Chitosan, a Biopolymer with triple action on postharvest decay of fruit and vegetables: Eliciting, antimicrobial and film-forming properties. *Front. Microbiol.* **2018**, *9*, 2745. [CrossRef] [PubMed]
54. Martínez, O.; Salmerón, J.; Epelde, L.; Vicente, M.S.; De Vega, C. Quality enhancement of smoked sea bass (*Dicentrarchus labrax*) fillets by adding resveratrol and coating with chitosan and alginate edible films. *Food Control* **2018**, *85*, 168–176. [CrossRef]
55. Thabet, M. Application of chitosan and oxalic acid combined with hot water to control postharvest decay of strawberry fruits caused by *Botrytis cinerea* and *Rhizopus stolonifer*. *Middle East J. Appl. Sci.* **2019**, *9*, 63–77.
56. Prusky, D.; Romanazzi, G. Induced resistance in fruit and vegetables: A host physiological response limiting postharvest disease development. *Annu. Rev. Phytopathol.* **2023**, *61*, 1. [CrossRef] [PubMed]

MDPI
St. Alban-Anlage 66
4052 Basel
Switzerland
www.mdpi.com

Coatings Editorial Office
E-mail: coatings@mdpi.com
www.mdpi.com/journal/coatings

www.ingramcontent.com/pod-product-compliance
Lightning Source LLC
Chambersburg PA
CBHW040716270926
51850CB00023B/2667
9783725806614